735

Jeremy Christian Short

from Great Grandfather Charles Wilcox

For all the men
who designed, built, overhauled,
serviced, and ran the big steam power . . .
and for those who envied them the experience,
especially K. J. M.

STEAM'S FINEST HOUR

EDITED BY DAVID P. MORGAN

MILWAUKEE 3, WISCONSIN
PRINTED IN U. S. A.

In West Virginia's New River Gorge Chesapeake & Ohio 2-6-6-6 No. 1624 struggles mightily with 90 loads of coal after being pulled from storage to spell diesels in 1956. Photo: Philip A. Weibler.

What is the "modern steam locomotive"?

IT is infrequent in history of any sort that the writer can pinpoint with the authority of Genesis' "In the beginning . . . ." The modern steam locomotive, in this instance, evolves in design according to its context in railroading; in 1910 "modern power" was the 2-8-0, in 1920 the 2-8-2, in 1930 the 4-8-4. In our time Norfolk & Western (as good a source as exists) declared that the term meant "a locomotive designed with a high-capacity boiler, equipped with roller bearings on all engine and tender wheels, one-piece cast steel bed frame, improved counterbalancing, and complete mechanical and pressure lubrication." To be sure, such an engine was an end result of the process of evolution. Rail colossus E. H. Harriman declared in the early years of the century, for example, that locomotives had attained their maximum practical size because they had reached the endurance of the human fireman — and barring the invention of the mechanical stoker he would have been right. But assuming such components kept pace, can the latter-day historian isolate an era, even an engine, and define it as "the beginning"?

To a degree, yes. Almost all elements of N&W's definition of the modern steam locomotive can be traced back into the 1920's, for example; and it is at least reasonable to argue that the line of demarcation between old and new was drawn on the erecting floor of the Lima Locomotive Works in the spring of 1922. For there a 2-8-2 was taking shape to the design of W. E. Woodard and on the budget of the builder itself. Not a bigger Mikado (No. 8000 weighed less than 2 per cent more than the Michigan Central 2-8-2's she was being built to compete against) but a far better one. Other experimental probings of steam power capacity had leaned more on size than science to make their point. Only six years before Baldwin had engineered a 2-8-8-8-4 for Virginian (with just 75 square feet of grate area!) and a designer had seriously drawn blueprints for a Quintuplex — a two-cab 2-8-8-8-8-8-2! Woodard, though, was not content with existing 2-8-2 and 2-10-2 designs, much less articulateds. He concentrated on boiler capacity on the assumption that a booster could help overcome starting resistance but only ample firebox size would permit sustained horsepower production at 30 to 35 miles per hour without excessive firing rates.

The 8000 outperformed existing and just-as-heavy New York Central System 2-8-2's so easily that the road created its famous H-10 class on the basis of the Lima prototype, and bought more than 300 such Mikes within just two years. Thus encouraged, Lima developed its A-1 in 1925, a locomotive that originated the 2-8-4 wheel arrangement in order to place design emphasis squarely (and forever as it turned out) on steam supply and horsepower. No other steam locomotive built after World War I had such a remarkable, lasting, and beneficial effect upon the American scene. Simultaneous experimentation by other builders and roads on high-pressure boilers, compounding, and 3-cylinder transmissions all came to nought. But the free-steaming A-1 preached a sermon which was taken to heart. "Modern" no longer meant the addition of another driving axle and more adhesive weight; modern meant horsepower — *i.e.*, power at speed. Indeed, development proceeded so rapidly thereafter that Mopac, for instance, was rebuilding a series of 10-year-old Woodard 2-8-4's into 4-8-4's by 1940 — not because they couldn't steam but because their 63-inch drivers couldn't roll the hotshots fast enough.

Two years after the A-1 another engine was spreading the same gospel in the field of high-speed passenger service. New York Central's pioneer Hudson No. 5200 of 1927 weighed only marginally more than the system's K-5 Pacific (and actually less on drivers), yet far exceeded the heavy 4-6-2 in all areas: indicated cylinder horsepower, over-all thermal efficiency, power-to-weight ratio. On test the 5200 handled up to 26 coaches at 75 miles per hour, and on one unforgettable occasion she was worked in full forward gear at 65 with barely a tremor on the boiler pressure gauge needle.

Those were exciting years — years of ferment, reappraisal, and challenge. As the 1920's raced toward an unforeseen financial crisis ideas were proposed, tested, adopted or discarded. Occasionally the idea was sound enough — more efficient steam distribution, for example — but the means were inadequate. For 30 years and more designers refined Baker and Walschaerts valve gears, yet agreed that their best efforts hampered the flow of steam in and out of valves and cylinders; and for as many years one alternative after another — from Caprotti to Uniflow to Franklin rotary cam poppet valves — was tried and found wanting under heavy-duty U. S. operating conditions. The designers were still searching when steam died.

"The best of locomotive art" — to quote Baldwin — was incorporated in the duplex-drive layout, a 4-cylinder but rigid-frame locomotive designed to lessen the high piston loads and crankpin maintenance of the larger 4-8-4's and 2-10-4's. The principle was excellent but a practical application of it proved elusive. Baltimore & Ohio's Mount Clare Shops constructed a

←In the spring of 1922 erecting floor of Lima held Mike of unorthodox design for the MC↓.

Lima A-1 2-8-4 with Milwaukee's Pioneer Limited *on a demonstration run in October 1925.*

prototype 4-4-4-4 with cylinders front and rear in 1937. Two years later all three major commercial builders pooled their talents on an 84-inch-drivered 6-4-4-6 with cylinders front and center which was built in Pennsy's Altoona Shops. Visitors to the New York World's Fair were urged to inspect "the world's largest, fastest, most powerful passenger steam locomotive," or simply, "the big engine." There was more to come as Pennsy experimented with a pair of 4-4-4-4's, then a 4-6-4-4, and finally mass-produced more than 75 standardized duplex-drives. Other roads as divergent as New York Central and Texas & Pacific expressed interest, and Baldwin — which had been quietly promoting the duplex-drive concept since the early days of the depression — was not above implying that Pennsy's yacht-nosed racers definitely dated the Northern and could perchance cramp the diesel. This was not to be. The duplex-drive began and died as an experiment, remembered more for being the reciprocating engine's final evolution than for the results obtained thereby.

Another idea caught fire, though. In 1930 the proposal of placing all axles of a steam locomotive on roller bearings was viewed so suspiciously that the Timken Roller Bearing Company couldn't even obtain permission to equip an existing engine at its own expense. So in 1930 Timken (and 52 suppliers who contributed components on an "open account" basis pending two years of tests) bought a demonstrator 4-8-4 from Alco. No. 1111, the famous *Four Aces*, proved as successful a salesman as railroading had ever seen . . . an engine that could rate newspaper coverage by virtue of being pulled by a trio of Pennsy office girls in Chicago, or wrestle with the tonnage of a 132-car coal train, or hit the summit of a 4-mile Mopac grade varying from 1.75 to 2 per cent at 21 miles per hour with an 18-car *Hot Springs Special* behind the tank. More important, the 1111 ran off 119,600 miles in two years without roller-bearing troubles of any kind on any axle. Her performance reflected increased starting power because of reduced bearing resistance; free coasting (and hence less tendency for the train behind to bunch at the tender drawbar); cool operation (in winter service frost actually collected on journal covers and axle hubs); reduced maintenance; and sharply boosted availability. Purchased by Northern Pacific and renumbered 2626, *Four Aces* survived until diesels and long after roller bearings on all axles were considered as necessary on a steam locomotive as the engineer's seatbox.

Other experiments failed and succeeded, too. For example, a long and tortuous trail of experimentation was required before the principle of articulation bridged the gap between the slow-moving compound 2-8-8-2 of World War I and the high-speed simple 4-6-6-4 of World War II. Three major problems had to be solved: (1) boiler capacity had to be sufficient to permit simultaneous high-pressure steam consumption by 4 cylinders; (2) steam connections from the boiler to the cylinders had to be tight and of reasonable length to prevent a severe temperature drop; and (3) articulation had to be of such design as to permit safe operation at higher speeds. Step by step, one failure after another groped toward a total solution. The Trans-Siberian had a 5-foot-gauge simple 2-4-4-0 as early as 1902 and Pennsy bought a simple 2-8-8-2 in 1912, followed it with an extremely powerful simple 2-8-8-0 (tractive effort: 135,000 pounds) in 1919. None of these engines was a success; but by the end of the 1920's the technique of using steam once aboard an articulated had been mastered to the degree that several roads were rebuilding older compounds into single-expansion 2-8-8-0's, and the simple 2-8-8-2 was a staple in the builders' catalogs. Chesapeake & Ohio turned to simple articulateds, for instance, simply because its Allegheny tunnels would not accommodate the low-pressure forward cylinders of larger compounds. Baltimore & Ohio tried to go the last mile in 1930 by installing a pair of 70-inch-drivered Baldwin simple 2-6-6-2's in mountain freight service. The watertube-fireboxed 7400 and the conventional-boilered 7450 were designed to replace Mikes and 2-10-2's. For a season the 7400 even appeared in passenger service as a 4-4-6-2. These KK-class engines were not repeated.

Five years later, however, the high-speed simple articulated was at work for Seaboard Air Line as a 69-inch-drivered 2-6-6-4, and a year later, in 1936, Union Pacific's inventive Otto Jabelman had taken the wraps off the first 4-6-6-4. Anatole Mallet, who lived long enough to vigorously protest the concept of single-expansion articulateds, would surely have been thunderstruck if he had come upon a Union Pacific 3900 rolling across the California desert at a mile a minute with a 22-car *Challenger* behind the tank.

Monumental progress (there is no other term for it) was made as a direct result of the experience derived from 8000, A-1, 5200, 1111, *et al.* Whereas the Pacific of 1920 could average 10,000 miles a month if pushed, the Hudson of 1940 could put away 20,000 miles without breathing hard. Spectacular gains were made even from one subclassification to the next of the same wheel arrangement. For example, the New York Central's 1927-design J-1 4-6-4's averaged 194,217 miles between classified repairs in 1941; in the same year the 1937-design J-3's averaged 232,768 miles. Consider also the experience of Union Pacific in two world wars. In 1918 1106 locomotives produced 30.4 billion gross ton-miles; in 1943 1117 engines ground out 85.2 billion gross ton-miles. For virtually the same number of locomotives to nearly triple the railroad's work output meant that (1) average engine tractive effort had soared (from 40,000 pounds in 1917 to 63,000 pounds in 1943); and (2) engine availability and utilization had kept pace. One may sympathize with the Washington planners to whom all steam engines looked more or less alike, the economists who in all good faith were fearful of a complete transport breakdown during the war.

Nowhere was the change more apparent than in the cab. No one, for example, admired Pennsy's J1 2-10-4 more than the crews who'd cut their teeth on the road's I1 2-10-0's. The ponderous Decapod possessed the riding qualities of an obstreperous Army mule, and its 70-square-foot grate allowed no margin of safety under sustained steaming conditions. A J1, on the other hand, had 121.7 square feet of grate, a ride like a Pullman, and a forgiving nature. "Put a little warm water in 'em and they'll go," was the testimonial of a Union Pacific man in behalf of the Big Boy 4-8-8-4's. Upon one occasion a New York Central Niagara arrived in Albany, N. Y., from Harmon with her fire in shreds and the pressure declining to 200 pounds because of a leaky flue and shoddy firing. Ten minutes' worth of work with the hook and a hand scoop renewed a fire hot enough to raise the pressure back to 270 pounds even as the 4-8-4 was climbing the 1.63 per cent grade and swinging to the 3½-degree curvature of Albany Hill. An observer could make readable notes standing on the deck of the same locomotive at 90 miles per hour. Or think of the stability of Milwaukee Road's streamlined, 7-foot-drivered Hudsons, one of which once snapped a rod at 100 and survived the ordeal to roll to a safe stop.

Throughout steam's finest hour the appearance of the machine steadily improved. Once upon a time esthetics seemed incompatible with either size or science. Primeval Mallets were often grotesque creatures obviously suffering from glandular ailments, and the pioneer Lima Super-Power engines looked like nothing so much as mobile laboratories. (Compare in the pages

When new in 1927, original New York Central Hudson, No. 5200, charges up Albany Hill.

Timken's Four Aces *with NP's* North Coast Limited *at Jamestown, N. Dak., in 1931.*

R. V. Nixon.

of this book, if you will, the 2-10-4's constructed for Texas & Pacific and for Kansas City Southern by the same builder but 12 years apart.) Two circumstances cleaned up external appearances. First, the locomotives grew so large that their auxiliaries from pumps to bells had to be molded into the over-all design in order to keep within inflexible clearances. And second, the designers often made a deliberate effort to smooth out the jacketing and harmoniously arrange the location of feedwater heaters, domes, and running boards. UP's 4-8-8-4, for example, had an engine weight of 762,000 pounds vs. 634,500 pounds for the road's largest 4-6-6-4, yet so compact was the bigger articulated that the two classes looked remarkably identical at a quick glance across a busy engine terminal. Again, it required a practiced eye indeed to differentiate between N&W's streamlined J and K2 engines when they were spotted at speed.

The challenge of streamlining machines weighing more than 200 tons produced many variations on a theme. The wind-tunnel school of designers came up with engines such as Milwaukee's *Hiawatha* Atlantics which were skirted down to their axles, but soon the realization dawned that streamlining's practical function was to lower customer rather than wind resistance, as one realist put it. Essentially, this amounted to applying a bullet, or shovel, nose and a so-called skyline casing over stack and domes — plus a few gallons of bright Duco. One of the most salable locomotives — Southern Pacific's GS-class *Daylight* power — was technically one of the least streamlined. This became apparent when diesels bumped the 4-8-4's into commutation work. The process of destreamlining a GS-4 amounted to little more than ripping off her running board skirts and painting her black. Even the stripped-down wartime GS models kept their skyline casings intact on grounds that their purpose was more to lift the smoke than to catch the eye.

All of these super steam locomotives, no matter how powerful or fast or efficient, eventually fell before the invincible diesel. Neither the devisings of the builders nor the inclinations of the coal roads could stalemate the advance. And yet perhaps it is our warmest recollections of the modern steam locomotive that may be traced to those hours in which she was on trial for her life . . . and even after the verdict had been handed in. When, say, was a Central Niagara in better form than during those

Baltimore & Ohio pioneered with articulated 7400.

ghastly winters of the late 1940's when the smallish steam generators of early diesel passenger units proved inadequate and the 6000's came back to reign in glory . . . even on Nos. 25 and 26? What of Union Pacific's Challengers and Big Boys, "stored serviceable" for year after year, only to be pressed back into mainline work during the late-summer traffic peak and there to perform so vigorously that their engineers frequently came upon the yellow blocks of diesel-powered trains ahead? Certainly one of the more memorable Rocky Mountain Railroad Club excursions occurred behind the tank of a Baldwin 4-8-4 which took the curves through the Royal Gorge at a pace which eliminated aisle traffic in the coaches . . . and surmounted the summit thereafter at Palmer Lake at a speed and with a trailing load that surprised even the Rio Grande.

The last steam stronghold — Norfolk & Western — staged as dramatic and telling a show of the American locomotive as the continent afforded. The creations of Roanoke Shops literally defied the diesel to come and get 'em, and year after postwar year the A's and J's and Y6's contrived to break the records they themselves had established. The 70-inch-drivered simple 2-6-6-4 offers perhaps the classic example. This A-class engine was rated at 13,000 tons west across the 112-mile Scioto Division district between Williamson, W. Va., and Portsmouth, O., which includes a ruling grade of 0.3 per cent. In 1951 the railway added an auxiliary water tender to the 2-6-6-4 on the theory that the power previously expended starting up after an intermediate water stop could be used to keep a larger train rolling faster. On test the A proved capable of running the 112 miles nonstop in less than 4 hours with 14,500 tons. Moreover, the 2-6-6-4 burned only 3 per cent more coal and actually consumed 8 per cent less water while she was thus boosting gross ton-miles per train-hour by a sizable 32 per cent. For steam, the hour was late but never finer. I

EDITOR'S NOTE: The reader is advised that the specifications included for each engine refer to the subject or class represented in the builder photo *at the time of construction* and do *not* account for subsequent renumberings or mechanical revisions. All of the photographers whose credit lines appear in this book share the credit for its pictorial scope with those unfortunately anonymous craftsmen responsible for the builder prints. Special thanks are due H. L. Broadbelt, Everett L. DeGolyer Jr., Elwin K. Heath, R. H. Kindig, William W. Kratville, George Sennhauser, Jim Scribbins, and H. K. Vollrath for rendering assistance in an hour of editorial crisis; and these roads aided and abetted our cause: CGW, GN, IC, KCS, MP, P&LE, WM, and WP. Art Director David A. Strassman designed the book; George A. Gloff created the cover artwork; the three R's — Gil Reid, Russ Richardson, and Ted Rose — handled layout and pasteup; and A. L. Schmidt did retouching. Rosemary Entringer was charged with copy preparation. Publisher A. C. Kalmbach endorsed and encouraged the work, more out of an affection for steam than because of business instincts, the editor suspects. Finally, Rand McNally was our understanding printer. Space will not permit personal credit to all the other parties whose work stands behind the by-line and publisher's imprint of this book. The selection of photography, choice of engines, statistical accuracy, and expression of opinion in *Steam's Finest Hour* are the editor's responsibility; any failings in these areas are his alone. — D.P.M.

CONTENTS

Arthur F. Knauer.

NEW ENGLAND

. . . where the seeds of Super Power fell upon rocky soil and flourished

THE abundant life for New England railroading died well in advance of the depression. Once well-to-do planters, the regional roads — too many of them in too few square miles — were cast into the lot of sharecroppers, grubbing out an uncertain existence in rocky, impoverished soil. It was not an expected or happy turn of events. For all of New England's maritime instincts, it had possessed model railway plants at home and had loomed large elsewhere wherever rails were spiked home in the Western Hemisphere. The disposition of Boston bankers once could have sent the 4-4-0 across a given frontier or left it to the prairie dogs. Native sons from Dodge to Meiggs to Willard had been in demand for management from Pennsylvania to Peru. And wherever men laid track there had been the Taunton locomotive, the Osgood Bradley car, the Howe truss.

A notable, proud tradition. But such a historical value was of small comfort as New England came to grips with a domestic crisis a half century later. For as rail revenues stabilized, then receded in the 1920's, the region was exposed to circumstances that would have tried the patience of a Job. The topography and scope of the area implied helper districts, short hauls, and high terminal expenses. A once lucrative milk and passenger business became increasingly vulnerable to the paved road. And the elements, never predictable down East, seemed to take a peculiar delight in assaulting the weakened railroads with flood and hurricane and snow.

On the last lap into North Station, Boston, Pacific Kwasind *of Boston & Maine, Lima P-4-b 3715, whips through Belmont, Mass., at eventide in May 1947.*

The depression tightened the vise. Boston & Maine (with Maine Central hanging on its back for management after 1932) had submitted to that stomach pumping known politely as "voluntary reorganization" as early as 1926 but still weathered the storm only by virtue of diplomatic summit meetings with the R.F.C. Boston & Albany and Central Vermont were bailed out by their parents, New York Central and Canadian National, respectively. New Haven and Rutland both foundered on the rocks. Only the isolated, top-of-Maine Bangor & Aroostook, which managed to sustain itself on a diet of pulp and potatoes, never had occasion to fear the sheriff.

The application of modern steam power in New England was spotty and spasmodic. Paradoxically, the region became at once the birthplace of Super-Power and a U.S.R.A. holdout. New Haven, the largest, confined its acknowledgment of the new era to a token purchase of 10 4-6-4's. In common with onetime stockholder Pennsy, NYNH&H seemed preoccupied with its electrification, and except for a flyer into three-cylinder switchers and 4-8-2's of mixed blessings, the road managed to hold the fort until dieselization with essentially World War I engines. Talk of duplex-drives never reached the contract stage. The appearance of modernity which Elesco feedwater heaters lent to fat Santa Fe's on the Maybrook haul and I-4 Pacifics on the *Merchants* was betrayed by the dates on their data plates. Appearances could and did deceive!

Boston & Maine took the opposite tack. With approximately half the revenue of New Haven to support as many route-miles, B&M went through two motive-power metamorphoses to reduce its operating expenses in tortuous, heavily graded territory. In 1928-1929 the road acquired 25 T-1 2-8-4's which were virtual duplicates of Lima's famous barnstorming A-1. Unorthodox, hooded, chunky machines, the Berks were placed in pools out of Mechanicville, N. Y., to Boston and Portland. Easier on the coal pile than B&M's older 2-10-2's (a fundamental consideration in a region which had to import all its fuel), the T-1's also trimmed hours off symbol freight schedules designed to attract competitive bridge traffic. The second reformation took place when B&M sought a dual-purpose engine which could alternate between tonnage and the increasingly heavy summer-season passenger trains such as the *East Wind*. The T-1, originally billed as available for varnish, was hampered by a 63-inch wheel, and the 10 outrageously handsome Pacifics which arrived from Lima in 1934 and 1937 were too ladylike for much beyond merchandise service. Baldwin resolved the issue with the heavier R-1 Mountain. These 4-8-2's, of which B&M eventually purchased 18 in various subclasses, could haul almost the tonnage of a T and more than made up for the slight deficiency by raising the maximum speed limit from 45 to 70 miles per hour, thanks to 73-inch drivers. Incidentally, the R-1-d is illustrated in this volume on page 27 by a duplicate series bought by Lehigh & Hudson River after the War Production Board froze engine design.

B&M's R-1 and T-1 engines illustrate, incidentally, the difference a decade can make in external appearance without a change in the fundamental quest for efficiency.

The competitive Boston & Albany gave the 2-8-4 its name, of course, and was so enthralled by the Lima A-1 that it added 55 to its roster and supplemented them with 20 Hudsons — which were a 75-inch-drivered modification of Central's famous J-1. At that point B&A steam power development lapsed, perhaps because Central by terms of its lease was required to pay 8 per cent per annum on the stock of a subsidiary which ceased to have a net income in 1931. What might be termed the second phase of modernism did occur on B&A to a limited degree after the war when the 3000- and 3100-series Mohawks ventured east to South Station, Boston, but their employment was more or less a stopgap until dieselization.

The balance of new steam locomotives purchased for New England service were of minor national significance. The most intriguing, hands down, were 10 Texas types on Central Vermont which achieved a press all out of proportion to their net worth by virtue of being "the largest freight locomotives in service on any New England railway." The 700's clearly looked the part, what with their Elesco heaters, all-weather vestibule cabs, and 12-wheel Vanderbilt tanks. Their specs, though, indicated more than a family resemblance to parent CNR's T-2-a 2-10-2's of 1924. CV's 2-10-4's were the lightest ever built (excepting that one experimental Alco shipped to Russia and certain meter-gauge Brazilians) and were more a formula for holding down axle loadings without sacrificing too much tractive effort than true Texases. And their 60-inch drivers held them to a maximum of 35 miles per hour. Nevertheless, they effectively eclipsed their 2-8-0 contemporaries on the CV roster, and they had the satisfaction of steaming down the Winooski and Connecticut rivers years after their B&A and B&M 2-8-4 neighbors had been reduced to scrap.

Bangor & Aroostook, which had the credit to buy New England's biggest power, never required it. The spud hauler, whose operating ratio seldom exceeded 70 per cent when other U. S. roads were in the courts en masse, bought its new steam engines in modest lots of two and four. There was never a true Super-Power locomotive on the property. BAR needed instead smaller power to gather up the loads at the potato sheds, assemble them into trains, and then cart these down to North Maine Junction for delivery to the Maine Central. It was work which could be expeditiously handled by 2-8-0's and light 4-8-2's costing $75,000 and $85,000 each, respectively. The fact that the Consolidation had been written off prior to World War I or that BAR's Mountain exerted less than 50,000 pounds tractive effort less booster concerned the 600-mile road not at all. Their portraits in latter-day editions of the *Locomotive Cyclopedia* must have warmed the heart of many an old master mechanic.

David Plowden.

Maine Central managed to operate from 1924 to dieselization without approaching the builders for new power, excepting a pair of light 4-6-4's acquired in 1930. Even the Hudsons remain an inexplicable purchase to outlanders. And then there's Rutland, so far behind the others in motive power that it was obliged to streamstyle a 1911 2-8-0 with Stephenson gear when its *Whippet* hotshot was introduced in 1938. Somehow the road's old power hung together until Alco could be contacted for construction of four economy-model 4-8-2's in 1946. Less than 10 years later the property had been dieselized and the Mountains were up for sale.

Thus it was that in New England the modern steam locomotive encountered enthusiastic acceptance and almost blunt rejection.

Leo Litwin.

DUAL-PURPOSE ELEGANCE BY LIMA

Boston & Maine P-4-a Pacific 3712 rolling the northbound *Ambassador* along the Merrimack River on New Hampshire Division rails at Tyngsboro, Mass., typifies 10 handsome 4-6-2's Lima produced for optional heavy passenger and fast merchandise service in 1934 and 1937. The P-4's, among the first new engines after the virtual abandonment of locomotive construction in the depression, were all named by school children.

FIRST PASSENGER POWER SINCE 1916

"Dollar for dollar," Baldwin said of New Haven's first 1400-series 4-6-4 in 1937, she could "show her heels" to any other form of motive power. Known as the Shore Line type (instead of Hudson), the I-5 was designed to haul 15-car limiteds between Boston and New Haven, Conn., relieving the strain theretofore levied upon rebuilt World War I-era I-4 Pacifics. The 1409 moves the big 16-car *Colonial* out of Saybrook, Conn.

T. J. Donahue.

Collection of George Sennhauser.

BOSTON & MAINE P-4-b PACIFIC (4-6-2) FOR PASSENGER SERVICE

Series	3715-3719	Weight on lead truck	61,500 lbs.
Cylinders, diameter and stroke	23 x 28 in.	Weight on trailing truck	68,500 lbs.
Driving wheel diameter	80 in.	Weight of total engine	339,800 lbs.
Boiler pressure	260 lbs. per sq. in.	Weight of tender, loaded	240,800 lbs.
Grate area	66.9 sq. ft.	Tender capacity	18 tons; 12,000 gals.
Evaporative heating surface	3848 sq. ft.	Tractive force, engine	40,900 lbs.
Superheater heating surface	966 sq. ft.	Tractive force, booster	11,900 lbs.
Weight on drivers	209,800 lbs.	Builder, date	Lima, 1937

NEW YORK, NEW HAVEN & HARTFORD I-5 SHORE LINE (4-6-4) FOR PASSENGER SERVICE

Series	1400-1409	Weight on lead truck	71,500 lbs.
Cylinders, diameter and stroke	22 x 30 in.	Weight on trailing truck	100,800 lbs.
Driving wheel diameter	80 in.	Weight of total engine	365,300 lbs.
Boiler pressure	285 lbs. per sq. in.	Weight of tender, loaded	332,000 lbs.
Grate area	77.1 sq. ft.	Tender capacity	16 tons; 18,000 gals.
Evaporative heating surface	3815 sq. ft.	Tractive force	44,000 lbs.
Superheater heating surface	1042 sq. ft.	Builder, date	Baldwin, 1937
Weight on drivers	193,000 lbs.		

Collection of H. L. Broadbelt.

TOO FEW, TOO LATE

The only new locomotives which Rutland acquired after World War I were 6 Pacifics in the 1920's. The road somehow managed to get its tonnage over an undulating Vermont profile with ancient Consolidations and light U.S.R.A. Mikes until 1945 when 4 brand-new 73-inch-drivered Alco Mountains went to work. No. 92 is drifting down into Burlington with time freight in 1946.

Philip R. Hastings.

LIGHT, NEAT, AND NEW

Sprightly 4-8-2 No. 108, bearing down on the depot at Millinocket, Me., with a work extra, typifies the modern power of Bangor & Aroostook — the road whose cheerful balance sheet permitted modest motive power purchases during the depression. BAR bought individualistic power with outboard engine-truck journals, front-end throttles, and horizontal-barred pilots — like Pennsy's.

Ray F. Higgins.

BANGOR & AROOSTOOK M MOUNTAIN (4-8-2) FOR FREIGHT SERVICE

Series	107-108	Weight on lead truck	50,000 lbs.
Cylinders, diameter and stroke	22½ x 30 in.	Weight on trailing truck	54,000 lbs.
Driving wheel diameter	63 in.	Weight of total engine	318,500 lbs.
Boiler pressure	240 lbs. per sq. in.	Weight of tender	233,000 lbs.
Grate area	65.6 sq. ft.	Tender capacity	18 tons; 12,500 gals.
Evaporative heating surface	3583 sq. ft.	Tractive force, engine	49,200 lbs.
Superheater heating surface	932 sq. ft.	Tractive force, booster	11,000 lbs.
Weight on drivers	214,500 lbs.	Builder, date	American, 1935

RUTLAND L-1 MOUNTAIN (4-8-2) FOR FREIGHT SERVICE

Series	90-93	Weight on lead truck	62,000 lbs.
Cylinders, diameter and stroke	26 x 30 in.	Weight on trailing truck	54,000 lbs.
Driving wheel diameter	73 in.	Weight of total engine	348,000 lbs.
Boiler pressure	230 lbs. per sq. in.	Weight of tender, loaded	207,770 lbs.
Grate area	67 sq. ft.	Tender capacity	14 tons; 11,000 gals.
Evaporative heating surface	3919 sq. ft.	Tractive force	54,300 lbs.
Superheater heating surface	1152 sq. ft.	Builder, date	American, 1946
Weight on drivers	232,000 lbs.		

Rail Photo Service: G. C. Corey.

LARGEST ENGINES IN NEW ENGLAND

Central Vermont's beetle-browed 2-10-4's employed their many axles more to spread engine weight (and they were the biggest steam power in New England) than to produce high horsepower. The Alcos were really more of an elongated 2-10-2 than a Super-Power machine. Nevertheless, the 35-mile-per-hour T-3-a's proved themselves as dependable as the Vermont winters they fought. The 707 nears Brattleboro, Vt., on B&M rails hugging the Connecticut River.

UNORTHODOXY IN THE BAY STATE

Boston & Maine 2-8-4 4009 blasting through South Ashburnham, Mass., explains pictorially how the early Lima Super-Power engines dispensed with esthetics in their quest for high efficiency. To comply with restrictive clearances this T-1-a mounts pumps, headlight, and bell on the pilot beam and carries a squared-off sandbox and a naked Coffin feedwater heater. What was lost in appearance must have been made up by their unmatched maintenance accessibility.

H. F. Harvey.

Collection of George Sennhauser.

BOSTON & MAINE T-1-b BERKSHIRE (2-8-4) FOR FREIGHT SERVICE

Series	4020-4024	Weight on lead truck	39,400 lbs.
Cylinders, diameter and stroke	28 x 30 in.	Weight on trailing truck	107,800 lbs.
Driving wheel diameter	63 in.	Weight of total engine	403,000 lbs.
Boiler pressure	240 lbs. per sq. in.	Weight of tender, loaded	330,600 lbs.
Grate area	100 sq. ft.	Tender capacity	23 tons; 17,500 gals.
Evaporative heating surface	5131 sq. ft.	Tractive force, engine	69,400 lbs.
Superheater heating surface	2111 sq. ft.	Tractive force, booster	12,000 lbs.
Weight on drivers	255,800 lbs.	Builder, date	Lima, 1929

CENTRAL VERMONT T-3-a TEXAS (2-10-4) FOR FREIGHT SERVICE

Series	700-709	Weight on lead truck	35,000 lbs.
Cylinders, diameter and stroke	27 x 32 in.	Weight on trailing truck	99,000 lbs.
Driving wheel diameter	60 in.	Weight of total engine	419,000 lbs.
Boiler pressure	250 lbs. per sq. in.	Weight of tender	269,800 lbs.
Grate area	84.4 sq. ft.	Tender capacity	20 tons; 13,500 gals.
Evaporative heating surface	4703 sq. ft.	Tractive force, engine	76,800 lbs.
Superheater heating surface	2208 sq. ft.	Tractive force, booster	13,100 lbs.
Weight on drivers	285,000 lbs.	Builder, date	American, 1928

Collection of Elwin K. Heath.

Wayne Brumbaugh.

THE EAST

. . . *where restricted clearances frequently*

IN utter contrast to diesel practice, the steam locomotive's development and application reflected a high degree of individual taste. How else does one explain the fact that the two largest railroads in the East adopted diametrically opposed motive power policies, yet neither's locomotives were emulated to any degree by other lines in the region? Pennsylvania, never a conformist, produced its Belpaire-boilered Pacifics and drag engines right up to the brink of a Super-Power era which argued for emphasis on horsepower instead of tractive force, as well as on dual-purpose locomotives. Extension of its electrification released more steam engines than Pennsy could find work for during the depression, so the Altoona test plant gathered moss. World War II produced at once a desperate need for new power and a W.P.B. clamp on custom design. Pennsy sampled a Norfolk & Western 2-6-6-4, but perhaps recalling an earlier and unhappy flirtation with jointed engines, the road settled for 125 2-10-4's of Chesapeake & Ohio T-1 pattern. The Texas J1's were eminently satisfactory, but the railroad's disinclination to buy anybody else's ideas was expressed in a rash of duplex-drive 4-4-4-4 and 4-4-6-4 engines after the war. Their theory was admirable but their performance hastened the day of dieseldom.

New York Central rostered literally hun-

In the glorious setting for which she was named, handsome Hudson J-1 5223 thunders by the Hudson River palisades with New York Central train No. 22.

limited engine size but never imaginative design

dreds of engine subclassifications in contrast to the Spartan simplicity of Pennsy's ranks. As late as 1946 Central owned no less than 38 minor or major variations on the 4-6-2 type, for example. After trying but not mass-producing a heavy Pacific and tabulating indifferent tests of a three-cylinder 4-8-2, Central was in a mood to explore the possibilities of four-wheel trailing trucks. Offspring Boston & Albany got a fleet of 2-8-4's following the 1925 test of Lima's A-1, then in 1927 Alco delivered J-1a 4-6-4 No. 5200 — first of her wheel arrangement and the prototype of an enormously famous fleet of 275 Hudsons. The handsome locomotive's ability to eliminate doubleheading by sustained high speed with heavy trains gave the 4-6-4 an undisputed reign over Central passenger affairs until 1940. In that year a parallel development of the 4-8-2 culminated in a fleet of dual-purpose L-3 Mohawks. These 4-6-4 and 4-8-2 wheel arrangements were finally and successfully merged in purpose in 1945 with the advent of the Niagara 4-8-4 — an engine whose longevity was abruptly cut short by diesels.

As interesting as these two rosters were in themselves, they cannot be said to have had a profound influence upon other roads — in the East or elsewhere. There was no more enthusiasm for Pennsy's duplex drives than there had been for its Belpaire boilers. And while Central was applauded for its exploitation of free-steaming, high-horsepower boilers, its engines were limited in size by clearance and axle-loading handicaps which did not apply elsewhere. After Central's, the next largest fleet of 4-6-4's in the U. S. was that of the Milwaukee Road — which numbered 27. And the 4-8-2 was never built in appreciable quantities for anyone but Central after 1930.

The universal engine for the other roads was the 4-8-4. Delaware & Hudson, Lackawanna, Lehigh Valley, Reading, and Western Maryland all enjoyed great success with the type, and both D&H and WM paired it with the 4-6-6-4 to produce the team which was so popular west of the Mississippi. The shift to the Northern was particularly symbolic of the shift in steam science on the D&H and Reading; both had previously depended upon low-drivered 2-8-0's.

Erie, always an interesting road to watch, fortified itself in the late '20's with the largest roster of Berkshires in the land, and these engines were easily able to sustain the road until dieselization (rebuilt heavy Pacifics were adequate for a marginal passenger service). It is reputed, incidentally, that Erie was talking to Lima about a 4-10-4 until the diesel intervened. The 2-8-4 itself got its final if extremely brief fling in the East when Central affiliate Pittsburgh & Lake Erie bought seven ferocious-looking but modestly proportioned Berkshires in 1948. They were the last new steam locomotives built by the American Locomotive Company.

Two other roads in the Pittsburgh camp took divergent paths. U. S. Steel's Bessemer & Lake Erie placed a low price tag on the questionable prestige of custom design and was happy to settle on a near duplicate of Burlington's M-4 as the prototype for its fleet of 47 2-10-4's (Q itself only had 28 such engines). Little 132-mile Pittsburgh & West Virginia was something else again. To handle heavy coal trains between Connellsville and Rook, Pa., 35½ miles, P&WV wanted to replace two of its old 2-8-0's or 2-8-2's with a single engine of 96,000 pounds or so tractive force — equivalent to that of a B&LE 2-10-4. But P&WV was tied to an axle loading of 65,000 pounds and the restrictions of 18-degree curves. So with Baldwin it pioneered the 2-6-6-4 wheel arrangement and, extraordinarily enough, specified a Belpaire boiler (the only articulated with one since Great Northern's homemade 2-8-8-2's).

Which summarizes the East . . . except for Baltimore & Ohio. B&O confounds the locomotive historian, for its roster contained a dazzling array of "firsts" and experiments which in net effect were negligible. Two examples will suffice. First, in 1930 B&O bought a pair of 70-inch-drivered 2-6-6-2's which, despite the age of their wheel arrangement, clearly forecast the possibilities of high-speed single-expansion articulateds. Second, in 1937 B&O built a rigid-frame 4-4-4-4 passenger locomotive. The railroad pushed no further on either front. Indeed, for all practical purposes its motive power development stabilized in the 1920's with the 4-6-2 and 2-10-2. The EM-1 Yellowstones as well as the rebuilt and homemade T-3 Mountains were stopgap measures pending realization of an earlier decision to go all diesel. (A B&O man would cite his road's early decision to cease steam experimentation in favor of buying diesels as a point of pride, not pardon, of course.)

The dominant Eastern railroads, New York Central and Pennsylvania, fought the diesel down to the wire, partially because of their coal interests and partially because both no doubt resented the intrusion of a totally alien form of power. For many months after World War II both railroads staged a spectacular steam show. Central's great Niagaras reeled off 28,000 to 30,000 miles per month, pausing only once between Harmon, N. Y., and Chicago for coal and scooping water at up to 80 miles per hour. And Pennsy feverishly worked with every steam distribution system from Walschaerts to poppet to turbine in order to preserve the coal-burning locomotive. Those who witnessed the contest will never forget. I

Jim Shaughnessy.

J. P. Lamb Jr.

RACE HORSES OF CHAMPION STOCK

New York Central's super Hudsons of 1937 produced almost 25 per cent more drawbar pull at 70 miles per hour than their celebrated sire of 1927. A subfreezing December 1955 wind blows whistle steam past cab as J-3a 5439 with the *Queen City* yells for a Dayton (O.) crossing.

"MOUNTAIN" WAS INAPPROPRIATE . . .

In the twilight of Central steam power L-4a Mohawk 3106 leans into westbound tonnage at Lawrenceburg Junction, Ind. Older Alco Mohawk 3010 is the partner. Although Alco assisted in early developmental work on the dual-purpose 4-8-2 (a wheel arrangement the water-level NYC could hardly bring itself to call Mountain) and produced most of the Mohawks, the highest expression of the class came with the 50 L-4's turned out by the Lima Locomotive Works in 1943.

John Krave.

NEW YORK CENTRAL J-3a HUDSON (4-6-4) FOR PASSENGER SERVICE

Series	5405-5444	Weight on lead truck	66,000 lbs.
Cylinders, diameter and stroke	22½ x 29 in.	Weight on trailing truck	98,000 lbs.
Driving wheel diameter	79 in.	Weight of total engine	360,000 lbs.
Boiler pressure	275 lbs. per sq. in.	Weight of tender, loaded	314,300 lbs.
Grate area	82 sq. ft.	Tender capacity	30 tons; 14,000 gals.
Evaporative heating surface	4187 sq. ft.	Tractive force, engine	43,440 lbs.
Superheater heating surface	1745 sq. ft.	Tractive force, booster	12,100 lbs.
Weight on drivers	196,000 lbs.	Builder, date	American, 1938

NEW YORK CENTRAL L-4a MOHAWK (4-8-2) FOR DUAL SERVICE

Series	3100-3124	Weight on lead truck	73,100 lbs.
Cylinders, diameter and stroke	26 x 30 in.	Weight on trailing truck	58,400 lbs.
Driving wheel diameter	72 in.	Weight of total engine	397,300 lbs.
Boiler pressure	250 lbs. per sq. in.	Weight of tender, 2/3 load	305,700 lbs.
Grate area	75.3 sq. ft.	Tender capacity	42½ tons; 15,500 gals.
Evaporative heating surface	4676 sq. ft.	Tractive force, engine	59,900 lbs.
Superheater heating surface	2103 sq. ft.	Builder, date	Lima, 1942-1943
Weight on drivers	265,800 lbs.		

BOUGHT INSTEAD OF DIESELS

When World War II pinched off the supply of road diesels Baltimore & Ohio wanted, it settled instead for 20 Yellowstone articulateds — the largest steam power on its roster. EM-1 2-8-8-4 7605 belches smoke from twin stacks as she rolls hotshot 96 over the summit at Terra Alta, W. Va., following 8.9 miles of stiff 2.12 per cent on Cranberry grade in the Allegheny Mountains.

Rail Photo Service: B. F. Cutler.

William P. Price.

A CHALLENGER FOR THE ALLEGHENIES

Western Maryland turned from Mallets to heavy, 61-inch-drivered Decapods in the 1920's, returned to articulation in 1940-1941 in an effort to achieve a 50-mile-per-hour locomotive that could still lug on the 1.75 per cent ruling grade up the Alleghenies. Extra 1208 West nearing the summit at Deal, Pa., is propelled upgrade by one of 12 4-6-6-4's which were purchased from Baldwin.

WESTERN MARYLAND M-2 CHALLENGER (4-6-6-4) FOR FREIGHT SERVICE

Series	1201-1212	Weight on lead truck	81,500 lbs.
Cylinders, diameter and stroke	22 x 32 in. (4)	Weight on trailing truck	117,234 lbs.
Driving wheel diameter	69 in.	Weight of total engine	601,000 lbs.
Boiler pressure	250 lbs. per sq. in.	Weight of tender, 2/3 load	338,250 lbs.
Grate area	118.8 sq. ft.	Tender capacity	30 tons; 22,000 gals.
Evaporative heating surface	5770 sq. ft.	Tractive force, engine	95,500 lbs.
Superheater heating surface	1735 sq. ft.	Builder, date	Baldwin, 1940-1941
Weight on drivers	402,266 lbs.		

BALTIMORE & OHIO EM-1 YELLOWSTONE (2-8-8-4) FOR FREIGHT SERVICE

Series	7600-7629	Weight on lead truck	50,700 lbs.
Cylinders, diameter and stroke	24 x 32 in. (4)	Weight on trailing truck	93,000 lbs.
Driving wheel diameter	64 in.	Weight of total engine	628,700 lbs.
Boiler pressure	235 lbs. per sq. in.	Weight of tender, 2/3 load	304,000 lbs.
Grate area	117.5 sq. ft.	Tender capacity	25 tons; 22,000 gals.
Evaporative heating surface	5298 sq. ft.	Tractive force, engine	115,000 lbs.
Superheater heating surface	2118 sq. ft.	Builder, date	Baldwin, 1944-1945
Weight on drivers	485,000 lbs.		

W. R. Osborne.

FOR NAMESAKE MOUNTAINS, POCONOS

A pioneer 4-8-4 operator, Lackawanna introduced a 77-inch-drivered Pocono (as DL&W dubbed it) in 1927 for passenger service, was soon buying a 70-inch version for Hoboken-Buffalo freight runs, finally achieved a 74-inch dual-purpose machine. No. 1628, blackening the sky above her, pours her 71,600 pounds tractive force into tonnage on neatly ballasted iron. She was new in 1932.

Mike Runey.

A WYOMING WIDE OPEN

Rivals Lackawanna and Lehigh Valley, faced with the similar challenge of hotshot schedules superimposed upon mountain gradients, found the logical answer in the 4-8-4. DL&W bought 55 and LV took 37. Wyoming-type 5218, a T-2-b built by Alco in 1943, has about 100 cars behind her tank as she blasts westward along Susquehanna River on Lehigh's Wyoming Division.

DELAWARE, LACKAWANNA & WESTERN Q-3 POCONO (4-8-4) FOR FREIGHT SERVICE

Series	1621-1630	Weight on lead truck	74,000 lbs.
Cylinders, diameter and stroke	28 x 32 in.	Weight on trailing truck	84,000 lbs.
Driving wheel diameter	70 in.	Weight of total engine	429,000 lbs.
Boiler pressure	235 lbs. per sq. in.	Weight of tender	287,200 lbs.
Grate area	88.2 sq. ft.	Tender capacity	22 tons; 15,000 gals.
Evaporative heating surface	5445 sq. ft.	Tractive force, engine	71,600 lbs.
Superheater heating surface	2243 sq. ft.	Builder, date	American, 1932
Weight on drivers	271,000 lbs.		

LEHIGH VALLEY T-2-B WYOMING (4-8-4) FOR FREIGHT SERVICE

Series	5211-5220	Weight on lead truck	72,500 lbs.
Cylinders, diameter and stroke	26 x 32 in.	Weight on trailing truck	104,000 lbs.
Driving wheel diameter	70 in.	Weight of total engine	451,000 lbs.
Boiler pressure	255 lbs. per sq. in.	Weight of tender, 2/3 load	304,400 lbs.
Grate area	88.3 sq. ft.	Tender capacity	30 tons; 20,000 gals.
Evaporative heating surface	5376 sq. ft.	Tractive force, engine	66,700 lbs.
Superheater heating surface	2095 sq. ft.	Tractive force, booster	12,300 lbs.
Weight on drivers	274,500 lbs.	Builder, date	American, 1943

Donald W. Furler.

CASE OF BORROWED BLUEPRINTS

Engine evolution on the 97-mile, bridge-route Lehigh & Hudson River jumped from heavy drag-service 2-8-0's to high-horsepower 4-8-2's in 1944 when three duplicates of B&M's R-1d class were obtained from Baldwin. Smoke-deflectored No. 11 has symbol freight HO-6 in tow near Warwick, N. Y.; lack of smoke is commendable.

ERIE THRIVED ON BIG BERKS

Erie was tremendously impressed by the performance of Lima's Super-Power A-1 2-8-4. For higher speed driver size was altered from 63 to 70 inches. Then all three major builders were called upon to deliver 105 Berkshires before the buying spree was over. Sample S-4 3403 gets the 111 cars of hotshot 79 out of Hornell, N. Y.

Robert R. Malinoski.

LEHIGH & HUDSON RIVER MOUNTAIN (4-8-2) FOR FREIGHT SERVICE

Series	10-12	Weight on lead truck	79,627 lbs.
Cylinders, diameter and stroke	28 x 31 in.	Weight on trailing truck	66,457 lbs.
Driving wheel diameter	73 in.	Weight of total engine	415,200 lbs.
Boiler pressure	240 lbs. per sq. in.	Weight of tender, 2/3 load	320,600 lbs.
Grate area	79 sq. ft.	Tender capacity	21 tons; 23,000 gals.
Evaporative heating surface	4511 sq. ft.	Tractive force, engine	67,000 lbs.
Superheater heating surface	1887 sq. ft.	Builder, date	Baldwin, 1944
Weight on drivers	269,116 lbs.		

ERIE S-4 BERKSHIRE (2-8-4) FOR FREIGHT SERVICE

Series	3385-3404	Weight on lead truck	49,000 lbs.
Cylinders, diameter and stroke	28½ x 32 in.	Weight on trailing truck	133,300 lbs.
Driving wheel diameter	70 in.	Weight of total engine	468,800 lbs.
Boiler pressure	250 lbs. per sq. in.	Weight of tender, loaded	378,000 lbs.
Grate area	100 sq. ft.	Tender capacity	28 tons; 20,000 gals.
Evaporative heating surface	5697 sq. ft.	Tractive force, engine	72,000 lbs.
Superheater heating surface	2545 sq. ft.	Tractive force, booster	13,000 lbs.
Weight on drivers	286,500 lbs.	Builder, date	Lima, 1929

Jim Shaughnessy

U. S. STEEL AND Q'S BLUEPRINTS

The railroads owned by U.S. Steel have never given a fig for prestige so long as their owner's ore was hauled to the mills economically. Accordingly, Bessemer & Lake Erie's 47 2-10-4's were virtual duplicates of Burlington's M-4. Missabe 714, fighting up to Proctor, Minn., was bumped by diesels off B&LE.

WHAT, NO BELPAIRE BOILER?

Pennsylvania J1 6402 strides across B&O en route to Columbus, O., from Sandusky. The interlocking plant is at Attica Junction. When W.P.B. restrictions frustrated Pennsy's inventive energies in World War II, the road bought 125 2-10-4's of essentially C&O T-1 design—but with a raised headlight, of course.

John Krave

PENNSYLVANIA
J1 TEXAS (2-10-4) FOR FREIGHT SERVICE

Series	6401-6500
Cylinders, diameter and stroke	29 x 34 in.
Driving wheel diameter	69 in.
Boiler pressure	270 lbs. per sq. in.
Grate area	121.7 sq. ft.
Evaporative heating surface	6568 sq. ft.
Superheater heating surface	2930 sq. ft.
Weight on drivers	379,493 lbs.
Weight on lead truck	64,541 lbs.
Weight on trailing truck	131,846 lbs.
Weight of total engine	575,880 lbs.
Tender capacity	30 tons; 21,000 gals.
Tractive force, engine	95,100 lbs.
Tractive force, booster	15,000 lbs.
Builder, date	Altoona, 1943-1944

BESSEMER & LAKE ERIE H1A TEXAS (2-10-4) FOR FREIGHT SERVICE

Number	601
Cylinders, diameter and stroke	31 x 32 in.
Driving wheel diameter	64 in.
Boiler pressure	250 lbs. per sq. in.
Grate area	106.5 sq. ft.
Evaporative heating surface	5904 sq. ft.
Superheater heating surface	2487 sq. ft.
Weight on drivers	354,300 lbs.
Weight on lead truck	35,530 lbs.
Weight on trailing truck	112,800 lbs.
Weight of total engine	502,630 lbs.
Weight of tender	375,500 lbs.
Tender capacity	23 tons; 22,000 gals.
Tractive force, engine	96,700 lbs.
Tractive force, booster	13,235 lbs.
Builder, date	Baldwin, 1929

R. L. Long.

AMERICA'S NEWEST PACIFICS . . .

Wide-fireboxed Reading 4-6-2 No. 215, one of 10 G-3's built in company shops in 1948, pauses perhaps disconsolately in late afternoon at Audubon, N. J., with Pennsylvania-Reading Seashore Lines local No. 609. The skirted G-3's barely had their paint dry when diesels bumped them off proposed runs (*e.g.*, the *Wall Street*).

THE PRESIDENTS . . . RE-ELECTED

Baltimore & Ohio's version of its rival's K4—20 olive-green 4-6-2's built in 1927 and originally named for U. S. presidents—really came of age 20 years after birth when rebuilt and streamlined Nos. 5301-5304 began hauling the all-coach *Cincinnatian* from Baltimore to Cincinnati in 12½ hours . . . running 95.7 per cent O.T.

William P. Price.

Courtesy N.R.H.S. *Bulletin.*

READING G-3 PACIFIC (4-6-2) FOR PASSENGER SERVICE

Series	210-219	Weight on lead truck	63,222 lbs.
Cylinders, diameter and stroke	25 x 28 in.	Weight on trailing truck	69,044 lbs.
Driving wheel diameter	80 in.	Weight of total engine	329,450 lbs.
Boiler pressure	260 lbs. per sq. in.	Weight of tender, loaded	257,540 lbs.
Grate area	95 sq. ft.	Tender capacity	19 tons; 12,500 gals.
Evaporative heating surface	2983 sq. ft.	Tractive force, engine	48,340 lbs.
Superheater heating surface	770 sq. ft.	Builder, date	Reading, 1948
Weight on drivers	197,184 lbs.		

BALTIMORE & OHIO P-7d PACIFIC (4-6-2) FOR PASSENGER SERVICE

Series	5301-5304	Weight on lead truck	70,000 lbs.
Cylinders, diameter and stroke	27 x 28 in.	Weight on trailing truck	66,500 lbs.
Driving wheel diameter	80 in.	Weight of total engine	347,500 lbs.
Boiler pressure	250 lbs. per sq. in.	Weight of tender, loaded	366,000 lbs.
Grate area	70.3 sq. ft.	Tender capacity	25 tons; 20,000 gals.
Evaporative heating surface	3843 sq. ft.	Tractive force, engine	50,000 lbs.
Superheater heating surface	950 sq. ft.	Builder, date	Baldwin, 1927; rebuilt Mt. Clare, 1946
Weight on drivers	211,000 lbs.		

Rail Photo Service.

DUPLEX DRIVE AND POPPET VALVES

Yacht-nosed Pennsylvania 4-4-4-4 No. 5517 eases away from Englewood, Ill., en route to the East. The Baldwin and homemade duplex-drive engines were far advanced in theory but disappointing in actual practice. They were styled by Raymond Loewy and they could run as fast as the engineer's nerves permitted. But they were never, as billed, equal to a GG1 electric.

WHAT WOULD HAVE BEEN A 4-10-4

Pennsy's postwar Q2 employed the duplex-drive layout to ease the main-rod load and improve steam distribution on what otherwise would have been the 4-10-4 wheel arrangement. The rigid-frame 4-4-6-4's weighed about as much as the typical Challenger. Homemade No. 6187 climbs the .51 per cent ruling grade west from Fort Wayne, Ind., with 68 cars behind her tender.

D. Allen Bauer

PENNSYLVANIA T1 4-4-4-4 FOR PASSENGER SERVICE

Series	5500-5524
Cylinders, diameter and stroke	19¾ x 26 in. (4)
Driving wheel diameter	80 in.
Boiler pressure	300 lbs. per sq. in.
Grate area	92 sq. ft.
Evaporative heating surface	4218 sq. ft.
Superheater heating surface	1680 sq. ft.
Weight on drivers	279,910 lbs.
Weight on lead truck	99,570 lbs.
Weight on trailing truck	122,720 lbs.
Weight of total engine	502,200 lbs.
Weight of tender, loaded	442,500 lbs.
Tender capacity	42.6 tons; 19,200 gals.
Tractive force, engine	64,650 lbs.
Builder, date	Altoona, 1945

PENNSYLVANIA Q2 4-4-6-4 FOR FREIGHT SERVICE

Series	6175-6199
Cylinders, diameter and stroke	19¾ x 28 in. (front) 23¾ x 29 in. (rear)
Driving wheel diameter	69 in.
Boiler pressure	300 lbs. per sq. in.
Grate area	121.71 sq. ft.
Evaporative heating surface	6725 sq. ft.
Superheater heating surface	2930 sq. ft.
Weight on drivers	393,000 lbs.
Weight on lead truck	96,050 lbs.
Weight on trailing truck	130,050 lbs.
Weight of total engine	619,100 lbs.
Weight of tender, loaded	430,000 lbs.
Tender capacity	39.8 tons; 19,020 gals.
Tractive force, engine	100,800 lbs.
Tractive force, booster	15,000 lbs.
Builder, date	Altoona, 1946

John F. McBride.

ENGINES THAT FOUGHT THE DIESELS

In the fall of 1946 New York Central ran a controlled test of six Niagara engines in passenger service against an equal number of diesels, and the S-1's averaged up to 27,221 miles per month apiece. No. 6020, working her way up the Hudson with the *Advance Empire State Express*, typifies the power that covered itself with glory in its last-ditch fight against the diesel invader.

INSTEAD OF EXPERIMENTALS

Under President Leonor F. Loree the Delaware & Hudson ordered some remarkably efficient if slow and expensive compounds. A decade later D&H began buying fast, high-horsepower locomotives in the interest of lucrative bridge-route traffic. No. 301 — one of 15 elephant-eared K-class 4-8-4's — sends smoke boiling skyward as she works southbound tonnage out of Plattsburg, N. Y.

Jim Shaughnessy.

DELAWARE & HUDSON K-62 NORTHERN (4-8-4) FOR FREIGHT SERVICE

Series	300-314	Weight on lead truck	92,500 lbs.
Cylinders, diameter and stroke	24½ x 32 in.	Weight on trailing truck	107,500 lbs.
Driving wheel diameter	75 in.	Weight of total engine	470,000 lbs.
Boiler pressure	285 lbs. per sq. in.	Weight of tender, 2/3 load	298,000 lbs.
Grate area	96.2 sq. ft.	Tender capacity	25 tons; 20,000 gals.
Evaporative heating surface	4477 sq. ft.	Tractive force, engine	62,040 lbs.
Superheater heating surface	1473 sq. ft.	Builder, date	American, 1943
Weight on drivers	270,000 lbs.		

NEW YORK CENTRAL S-1a NIAGARA (4-8-4) FOR PASSENGER SERVICE

Number	6000	Weight on lead truck	91,000 lbs.
Cylinders, diameter and stroke	25 x 32 in.	Weight on trailing truck	105,000 lbs.
Driving wheel diameter	75 in.	Weight of total engine	471,000 lbs.
Boiler pressure	275 lbs. per sq. in.	Weight of tender, 2/3 load	337,400 lbs.
Grate area	100.1 sq. ft.	Tender capacity	46 tons; 18,000 gals.
Evaporative heating surface	4632 sq. ft.	Tractive force, engine	62,500 lbs.
Superheater heating surface	1977 sq. ft.	Builder, date	American, 1945
Weight on drivers	275,000 lbs.		

Ralph E. Hallock.

BELPAIRE BOILERED 2-6-6-4's

Pittsburgh & West Virginia pioneered the 2-6-6-4 wheel arrangement in 1935 to resolve the problems of reasonably light axle loadings and tight curvature, and in the process evolved a neat, compact articulated. Belpaire-boilered No. 1102 swings to a curve on rock-ballasted iron as she rolls an extra from Monessen to Connellsville, Pa.

Fred McLeod.

SCHENECTADY'S STEAM SIGN-OFF

Bald-faced Pittsburgh & Lake Erie A-2a Berkshires looked as if they had swallowed their stacks. After the delivery of series 9400-9406 in the summer of 1948 American Locomotive ceased all steam production. By 1955 P&LE had bumped the 9401 off its property and she was getting out of Greensburg, Ind., east with Big Four freight.

PITTSBURGH & LAKE ERIE
A-2a BERKSHIRE (2-8-4) FOR FREIGHT SERVICE

Series	9400-9406
Cylinders, diameter and stroke	26 x 32 in.
Driving wheel diameter	63 in.
Boiler pressure	230 lbs. per sq. in.
Grate area	90.3 sq. ft.
Evaporative heating surface	4276 sq. ft.
Superheater heating surface	1881 sq. ft.
Weight on drivers	280,000 lbs.
Weight on lead truck	38,000 lbs.
Weight on trailing truck	108,000 lbs.
Weight of total engine	426,000 lbs.
Weight of tender	352,780 lbs.
Tender capacity	22 tons; 20,000 gals.
Tractive force, engine	67,130 lbs.
Builder, date	American, 1948

PITTSBURGH & WEST VIRGINIA J-1 2-6-6-4 FOR FREIGHT SERVICE

Series	1100-1102
Cylinders, diameter and stroke	23 x 32 in. (4)
Driving wheel diameter	63 in.
Boiler pressure	225 lbs. per sq. in.
Grate area	102.3 sq. ft.
Evaporative heating surface	5914 sq. ft.
Superheater heating surface	1873 sq. ft.
Weight on drivers	397,300 lbs.
Weight on lead truck	27,540 lbs.
Weight on trailing truck	103,200 lbs.
Weight of total engine	528,040 lbs.
Weight of tender, loaded	377,600 lbs.
Tender capacity	20 tons; 20,000 gals.
Tractive force, engine	97,500 lbs.
Tractive force, booster	16,000 lbs.
Builder, date	Baldwin, 1935

Collection of H. L. Broadbelt.

UNUSUAL WIDE-FIREBOXED 4-8-4'S

The rear ends of Reading's 30 Northerns smacked of those on the same road's famous roster of Wootten fireboxed Mother Hubbards; in fact, the T-1's were designed to burn a mixture of 10 per cent anthracite and 90 per cent bituminous coal. On loan to Pennsy the 2112 shrugs off a pelting rain in Northumberland, Pa., gets EC-2's 118 loads under way down the Susquehanna for Enola.

Don Wood.

William P. Price.

"A BABY COULD FIRE 'EM"

Western Maryland hung onto steam power until the bitter end because of an allegiance to on-line coal operators. Twelve J1-class 4-8-4's, dubbed Potomacs, were constructed by Baldwin in 1947 — the same year that diesel cab units began infiltrating the lines. Though short-lived, the J1's were well liked by crews. The 1401 is in charge of eastbound coal on the Hagerstown Division.

READING T-1 NORTHERN (4-8-4) FOR FREIGHT SERVICE

Series	2100-2129
Cylinders, diameter and stroke	27 x 32 in.
Driving wheel diameter	70 in.
Boiler pressure	240 lbs. per sq. in.
Grate area	94.5 sq. ft.
Evaporative heating surface	4920 sq. ft.
Superheater heating surface	1214 sq. ft.
Weight on drivers	278,200 lbs.
Weight on lead truck	67,000 lbs.
Weight on trailing truck	96,100 lbs.
Weight of total engine	441,300 lbs.
Weight of tender, loaded	367,700 lbs.
Tender capacity	26 tons; 19,000 gals.
Tractive force, engine	68,000 lbs.
Tractive force, booster	11,100 lbs.
Builder, date	Reading, 1945-1947

WESTERN MARYLAND J-1 POTOMAC (4-8-4) FOR FREIGHT SERVICE

Series	1401-1412
Cylinders, diameter and stroke	26½ x 32 in.
Driving wheel diameter	69 in.
Boiler pressure	255 lbs. per sq. in.
Grate area	106.7 sq. ft.
Evaporative heating surface	4974 sq. ft.
Superheater heating surface	2170 sq. ft.
Weight on drivers	290,000 lbs.
Weight on lead truck	98,500 lbs.
Weight on trailing truck	118,000 lbs.
Weight of total engine	506,500 lbs.
Weight of tender, 2/3 load	340,500 lbs.
Tender capacity	30 tons; 22,000 gals.
Tractive force, engine	70,600 lbs.
Builder, date	Baldwin, 1947

Jim Shaughnessy.

On a wintry Illinois day engine inspectors tap the vitals of Burlington 4-8-4 5634 with their hammers prior to a run to haul tonnage in a diesel shortage.

THE MIDDLE WEST

. . . a stamping ground for dual-service Northerns, trim 2-8-2's, seven-foot-drivered streamliners

FOR many granger railroads there was but one choice among the many wheel arrangements available for modern steam power. If the Middle West is seldom afflicted with helper grades, then neither is it a pancake-profile proposition. Freight traffic is more seasonal and of lighter density than it is in, say, the Pocahontas coal country. And there once often existed a 50-50 ratio between passenger and freight train-miles. So in steam what else but the 4-8-4? A Northern could shoulder the tonnage of a big Mike, yet roll it 50 to 55 miles per hour without losing steam or kinking the rail. When power became tight and tourists thick, the same engine could keep an 18-car *Olympian* to time.

The largest single class of 4-8-4's — Rock Island's 65 5000's — was ordered by a granger road as early as 1929. Many more Northerns went to work in the Middle West, in small lots and large, homemade and commercially built. Soo, which couldn't afford a mistake, purchased just 4 Northerns (the only new road power the line purchased between 1926 and 1947!) and received its money's worth. Toledo, Peoria & Western, establishing a name for itself under controversial George McNear Jr., bought 5. And at the other end of the order book there was Milwaukee Road, which built a pair of experimental 4-8-4's in its West Milwaukee Shops but turned to Alco and Baldwin to build a sizable fleet of Northerns; and neighbor Burlington, which bought its first 5600's but completed the series in its West Burlington (Ia.) Shops, albeit with Baldwin boilers. The smoke of the Northern traveled far and fast across the cornfields, and whether the cinders denoted a ponderous North Western H getting a meat train out of Council Bluffs or a Grand Trunk U-3-a racing across Michigan with hotshot 490 the story was the same. Success. The 4-8-4, perhaps particularly the 74-inch-wheel engines in the granger country, brought railroading full circle — back to the age of the 4-4-0 which could handle passengers and produce indiscriminately. (It's interesting to note that the diesel had to repeat the cycle before it, too, was manufactured as an all-purpose unit.) Many a *Fast Mail* of many a railroad was entrusted to the 4-8-4. One night, October 17, 1944, Burlington coupled 82 cars of Christmas mail for servicemen behind an O5A!

But if the 4-8-4 dominated the Middle West, it could not monopolize the region. The diesel, for one thing, precluded that. Specifically, it was No. 9900, a 196-foot, 97½-ton articulated lightweight poetically named the *Zephyr* that annoyed the steam advocates. Ordinarily they would have dismissed Burlington's streamliner as a newfangled doodlebug, but there was no denying the public response to the breathtaking nonstop, dawn-to-dusk run from Denver to Chicago or to the aeroplanelike speeds of the silver shovel-nose. The battle lines were drawn when Q ordered *Twin Zephyrs* for the Chicago-Twin Cities run.

The immediate reply in 1935 was North

Western's *400* — modernized heavy Pacifics with refurbished standard cars. Which was a stopgap measure, even if superb showmanship. The Milwaukee Road took a scientific approach, decided to duplicate the *Zephyr's* speed and glamour in a nonarticulated steam-powered train of greater flexibility. The two engines that Alco delivered for the new *Hiawatha* in 1935 were Atlantic in name and type only (in fact, builder and owner called Nos. 1 and 2 the Milwaukee type). The oil-fired, 300-pounds-pressure A-class 4-4-2's weighed more than a light U.S.R.A. Pacific, and because their main rods drove on the first driving axle they boasted a longer engine wheelbase than many 4-6-2's. And how those 84-inch drivers could roll! For the first time in American steam railroading, 100-mile-per-hour-plus operation became routine, not a *999* feat — and remained so even long after the original six-car trains had been expanded.

The motion of the A's was soon seconded by shrouded Hudsons. Quite simply, what the 4-4-2 could do, the F-7 4-6-4 could do with substantially more weight behind the tender. A record exists, for instance, of one of them averaging in excess of 100 miles per hour for 62 miles with a 16-car, 780-ton train. Incidentally, a total of 21 4-6-4's with 84-inch drivers and otherwise similar specifications were purchased by Milwaukee, North Western, and Santa Fe in 1937-1938, and in these three Hudson designs the steam locomotive achieved a zenith of speed and capacity.

There were Middle West operators whose traffic circumstances were not complicated by the need to woo the fickle passenger, much less carry him 100 miles per hour. Operating in heavily competitive territory, both Nickel Plate and Pere Marquette standardized on a jointly designed 2-8-4 to cement redball reputations. The capable Limas, in a typical PM example, made the Detroit-Chicago haul an overnight schedule, averaging 37.6 miles per hour with more than 3100 tons in tow. As late as 1948 NKP, for its part, pitted its Berkshires against borrowed road diesels and declared that the difference in costs was less than paper thin.

Excepting these 2-8-4's, Lima produced little of significance for most roads headquartered in its native Midwest. The big engines that won the luster for the diamond data plate handled the *Daylight* along the Pacific Coast and hoisted coal over the Alleghenies. Closer to home, the typical Lima order was a single 2-8-2 for Akron, Canton & Youngstown or perhaps four Mikes for Detroit, Toledo & Ironton.

Baldwin built the biggest by virtue of its rapport with the U. S. Steel roads. The largest of these properties is the Duluth, Missabe & Iron Range. The Missabe is, to be sure, no granger; indeed, DM&IR is only technically a common carrier. Its function is to haul iron ore off the Mesabi Range in huge, heavy trains of 21½-foot, 70-ton hoppers, tote them to the docks on Lake Superior for dumping, then bring the empties back. The locomotives that made the difference in the war were 18 Yellowstones remarkably similar to Western Pacific's earlier 251-series 2-8-8-2's but incorporating a 4-wheel trailing truck. The measure of their capacity is the fact that the 2-8-8-4's were hauling more tonnage in the twilight of their existence — trains of over 18,000 tons! — than when they were brand new.

Another one-commodity railroad in the Middle West is Chicago & Illinois Midland, a 121-mile road whose major purpose is to originate coal for the electric generating plants of owner Commonwealth Edison. In view of this tonnage and a ruling grade of 1.64 per cent, the Midland stuck to 2-10-2's, buying the drag engines brand new as late as 1931, later acquiring them secondhand from Coast Line and Wabash. They were among the best maintained engines in the country and the sight and sound of one of them tripling Petersburg Hill with a 6000-ton coal train was a rare experience down in the Prairie State corn country.

Bruce Meyer.

Chicago Great Western was another nonconformist. In 1930-1931 the Corn Belt Route bought 36 Texas types from Baldwin and Lima. These 2-10-4's were virtually identical to the pioneer engines of that wheel arrangement turned out five years earlier for Texas & Pacific. In common with other early Super-Power, CGW's T's pounded the track at any speed to speak of, so much so that at 60 the drivers tended to lift from the rail. New disc center drivers and lightweight rodding, applied in the late 1930's, resolved the difficulty so satisfactorily that at 60 miles per hour a modernized T produced less dynamic augment than the original engine had at 20.

The appeal of the Middle West for the locomotive student was such gateway cities as St. Paul, Omaha, St. Louis, and Chicago which produced a potpourri of wide extremes. Chicago, for example, where at Union Station the brutish 4-8-4's stood at parallel bumper posts with Pennsy doubleheaded K4 Pacifics and, across 320 feet of concourse, one could espy a 4-4-2 so streamlined that the naive thought her a diesel. Or Council Bluffs, where the vast tonnage brought across the river by Union Pacific was hauled away to the east by the huge 4-8-4's of Milwaukee and North Western, the chunky rebuilt Paducah Mikes of Illinois Central. Or, say, New Haven, Ind., where rivals crossed the diamond which was flanged by Nickel Plate 2-8-4's and Wabash 4-8-2's. Nowhere else in the U. S. could one see so many methods of turning superheated steam into train-mile statistics. I

Henry J. McCord.

THE SUPERLATIVE A CLASS

Milwaukee Road revived the 4-4-2 type and gave the wheel arrangement its finest hour in 1935 when 7-foot-drivered oil-burners pioneered the *Hiawatha.* No. 3 shows her mettle in April 1948 as she talks it up on the grade out of Elgin, Ill., with a 9-car westbound *Midwest Hiawatha.* For an A, 100 miles per hour was accelerating.

FOR HEAVIER HI'S . . . HUDSONS

Overwhelming public endorsement of Milwaukee Road's *Hiawathas* produced an order to Alco for six heavy, high-drivered 4-6-4's. In addition to 120-mile-per-hour running on the *Hi's* the F7's managed, on occasion, to accelerate a 1905-ton train from 0 to 70 miles per hour in 12 miles. The 104 gets out of Milwaukee south with No. 24.

Jim Scribbins.

Collection of David A. Strassman.

CHICAGO, MILWAUKEE, ST. PAUL & PACIFIC A MILWAUKEE (4-4-2) FOR PASSENGER SERVICE

Series	3-4	Weight on lead truck	72,700 lbs.
Cylinders, diameter and stroke	19 x 28 in.	Weight on trailing truck	73,000 lbs.
Driving wheel diameter	84 in.	Weight of total engine	290,000 lbs.
Boiler pressure	300 lbs. per sq. in.	Weight of tender	266,300 lbs.
Grate area	69 sq. ft.	Tender capacity	4000 gals. oil; 13,000 gals. water
Evaporative heating surface	3245 sq. ft.		
Superheater heating surface	1029 sq. ft.	Tractive force, engine	30,700 lbs.
Weight on drivers	144,300 lbs.	Builder, date	American, 1937

CHICAGO, MILWAUKEE, ST. PAUL & PACIFIC F-7 HUDSON (4-6-4) FOR PASSENGER SERVICE

Series	100-105	Weight on lead truck	82,500 lbs.
Cylinders, diameter and stroke	23½ x 30 in.	Weight on trailing truck	116,500 lbs.
Driving wheel diameter	84 in.	Weight of total engine	415,000 lbs.
Boiler pressure	300 lbs. per sq. in.	Weight of tender, loaded	375,000 lbs.
Grate area	96.5 sq. ft.	Tender capacity	25 tons; 20,000 gals.
Evaporative heating surface	4166 sq. ft.	Tractive force, engine	50,300 lbs.
Superheater heating surface	1695 sq. ft.	Builder, date	American, 1938
Weight on drivers	216,000 lbs.		

Donald E. Smith.

Quentin De Math.

ANOTHER VAN SWERINGEN BERKSHIRE

If there was a locomotive symbol of the vast railroad empire erected by the Van Sweringen brothers it was the Lima 2-8-4. Erie, Nickel Plate, Mopac, Chessie — they all had 'em. Pere Marquette, a member of the team, purchased 39 Berkshires from Lima between 1937 and 1944. Before PM's merger with C&O (which sent some PM 2-8-4's into Virginia) N-1 1226 digs in at Grand Rapids.

A DOMINION TOUCH IN THE U. S.

Canadian National's U. S. subsidiary Grand Trunk Western reports to the I.C.C. and buys its power locally, but the parent's design influence was quite apparent in steam. Aside from the air-scoop smoke deflector, GTW's six U-4-b streamlined 4-8-4's were counterparts of CNR's U-4-a's. No. 6406 has the eastbound *Maple Leaf* in hand as she cants to a curve leaving Valparaiso, Ind.

GRAND TRUNK WESTERN
U-4-b NORTHERN (4-8-4) FOR PASSENGER SERVICE

Series	6405-6410
Cylinders, diameter and stroke	24 x 30 in.
Driving wheel diameter	77 in.
Boiler pressure	275 lbs. per sq. in.
Grate area	73.7 sq. ft.
Evaporative heating surface	3852 sq. ft.
Superheater heating surface	1530 sq. ft.
Weight on drivers	237,900 lbs.
Weight on lead truck	62,000 lbs.
Weight on trailing truck	82,800 lbs.
Weight of total engine	382,700 lbs.
Weight of tender, 2/3 load	226,140 lbs.
Tender capacity	20 tons; 14,300 gals.
Tractive force, engine	52,500 lbs.
Builder, date	Lima, 1938

PERE MARQUETTE N-1 BERKSHIRE (2-8-4) FOR FREIGHT SERVICE

Series	1216-1227
Cylinders, diameter and stroke	26 x 34 in.
Driving wheel diameter	69 in.
Boiler pressure	245 lbs. per sq. in.
Grate area	90.3 sq. ft.
Evaporative heating surface	4777 sq. ft.
Superheater heating surface	1932 sq. ft.
Weight on drivers	277,600 lbs.
Weight on lead truck	50,900 lbs.
Weight on trailing truck	114,000 lbs.
Weight of total engine	442,500 lbs.
Weight of tender, 2/3 load	284,800 lbs.
Tender capacity	22 tons; 22,000 gals.
Tractive force, engine	69,350 lbs.
Builder, date	Lima, 1941

Collection of M. E. Kern.

Bob Lorenz.

A BERK AT DAWN

As soon as the depression eased, Nickel Plate began building the fleet of 80 Berkshires which wrote the meaning into its justifiably fine reputation for fast freight service. In the twilight of the S engines' reign No. 747, a Lima graduate of 1944, is all smoke and glory as she blasts west from Bellevue, O. The engineer appears to have curve and grade and tonnage well in hand.

AN H OF AN ENGINE

Chicago & North Western paid $120,000 apiece for its 35 big H-class 4-8-4's (no small sum in 1929), proudly proclaimed them "the largest dual-service locomotives in the world." Among other things, they did weigh almost twice as much as any previous C&NW engine. Several were later modernized into H-1's. In wartime 1943 the 3017 rolls tonnage through West Geneva, Ill.

Henry J. McCord.

NEW YORK, CHICAGO & ST. LOUIS S-3 BERKSHIRE (2-8-4) FOR FREIGHT SERVICE

Series	770-779	Weight on lead truck	55,040 lbs.
Cylinders, diameter and stroke	25 x 34 in.	Weight on trailing truck	123,220 lbs.
Driving wheel diameter	69 in.	Weight of total engine	444,290 lbs.
Boiler pressure	245 lbs. per sq. in.	Weight of tender, 2/3 load	288,840 lbs.
Grate area	90.3 sq. ft.	Tender capacity	22 tons; 22,000 gals.
Evaporative heating surface	4772 sq. ft.	Tractive force, engine	64,100 lbs.
Superheater heating surface	1932 sq. ft.	Builder, date	Lima, 1944
Weight on drivers	266,030 lbs.		

CHICAGO & NORTH WESTERN H NORTHERN (4-8-4) FOR DUAL SERVICE

Series	3001-3035	Weight on lead truck	87,000 lbs.
Cylinders, diameter and stroke	27 x 32 in.	Weight on trailing truck	123,000 lbs.
Driving wheel diameter	76 in.	Weight of total engine	498,000 lbs.
Boiler pressure	250 lbs. per sq. in.	Weight of tender	320,000 lbs.
Grate area	100 sq. ft.	Tender capacity	20 tons; 18,000 gals.
Evaporative heating surface	5214 sq. ft.	Tractive force, engine	65,200 lbs.
Superheater heating surface	2357 sq. ft.	Tractive force, booster	11,300 lbs.
Weight on drivers	288,000 lbs.	Builder, date	Baldwin, 1929

Collection of H. L. Broadbelt.

THE BEST OF BURLINGTON

One of the best of many fine granger 4-8-4's was Burlington's O5, an engine that could romp with a limited or a wheat train. Baldwin built the first 8 in 1930, later delivered the boilers for 28 more completed in Q's West Burlington (Ia.) Shops in 1936-1940. A homemade O5 heads Extra 5633 West over IC diamond at Mendota, Ill.

HEAVY BUT VERY LIGHT OF STEP

For a Hudson that was heavier than Pennsy's M1 Mountain, Burlington's S4 was quite a racer since she more than once kept a *Zephyr* to time. The class contained 13 Baldwins and a homemade job. Handsome 3001 gets away from the coaling stage at Mendota, Ill., in 1958 with a 16-car Illini Club excursion loaded with admirers.

Ray W. Buhrmaster.

John Pickett.

Collection of H. L. Broadbelt.

CHICAGO, BURLINGTON & QUINCY S-4 HUDSON (4-6-4) FOR PASSENGER SERVICE

Series	3000-3011	Weight on lead truck	69,320 lbs.
Cylinders, diameter and stroke	25 x 28 in.	Weight on trailing truck	114,830 lbs.
Driving wheel diameter	78 in.	Weight of total engine	391,880 lbs.
Boiler pressure	250 lbs. per sq. in.	Weight of tender, loaded	326,020 lbs.
Grate area	87.9 sq. ft.	Tender capacity	24 tons; 15,000 gals.
Evaporative heating surface	4247 sq. ft.	Tractive force, engine	47,700 lbs.
Superheater heating surface	1830 sq. ft.	Tractive force, booster	11,700 lbs.
Weight on drivers	207,730 lbs.	Builder, date	Baldwin, 1930

CHICAGO, BURLINGTON & QUINCY O5A NORTHERN (4-8-4) FOR DUAL SERVICE

Series	5621-5635	Weight on lead truck	72,650 lbs.
Cylinders, diameter and stroke	28 x 30 in.	Weight on trailing truck	121,990 lbs.
Driving wheel diameter	74 in.	Weight of total engine	476,050 lbs.
Boiler pressure	250 lbs. per sq. in	Weight of tender, loaded	362,000 lbs.
Grate area	106.5 sq. ft.	Tender capacity	27 tons; 18,000 gals.
Evaporative heating surface	5225 sq. ft.	Tractive force, engine	67,500 lbs.
Superheater heating surface	2403 sq. ft.	Builder, date	West Burlington, 1938, 1940
Weight on drivers	281,410 lbs.		

James G. LaVake.

MR. SPRAGUE'S MIKADOS

Minneapolis & St. Louis' neighbors almost divided up half the bankrupt granger and abandoned the rest. But Receiver Lucian Sprague ushered the pallbearers out. A key phase of his overhaul was simultaneous junking of worn-out light power and rebuilding of the newest engines — 35 low-drivered 2-8-2's. Booster-equipped 628 highballs west at Glen Lake, Minn., back in 1947.

Henry J. McCord.

THE TEXAS IN GRANGER COUNTRY

Chicago Great Western jumped out of the Mikado class in 1930 with orders to Baldwin and Lima for 36 Texas engines. The tendency of their 63-inch drivers to pound the rail and jiggle the seatboxes at 50 miles per hour was resolved in later years with lightweight rodding and disc center drivers. T-1 861 has perishables behind the tank as she pounds through St. Charles, Ill.

MINNEAPOLIS & ST. LOUIS
MAC-2-61 MAC ARTHUR (2-8-2) FOR FREIGHT SERVICE

Number	619
Cylinders, diameter and stroke	24 x 30 in.
Driving wheel diameter	59 in.
Boiler pressure	200 lbs. per sq. in.
Grate area	55 sq. ft.
Evaporative heating surface	3052 sq. ft.
Weight on drivers	214,000 lbs.
Weight on lead truck	25,300 lbs.
Weight on trailing truck	45,800 lbs.
Weight of total engine	285,100 lbs.
Weight of tender	193,900 lbs.
Tender capacity	17 tons; 10,200 gals.
Tractive force, engine	49,800 lbs.
Tractive force, booster	11,500 lbs.
Builder, date	American, 1916; rebuilt Cedar Lake, 1942

CHICAGO GREAT WESTERN T-3 TEXAS (2-10-4) FOR FREIGHT SERVICE

Series	874-879
Cylinders, diameter and stroke	29 x 32 in.
Driving wheel diameter	63 in.
Boiler pressure	255 lbs. per sq. in.
Grate area	100 sq. ft.
Evaporative heating surface	4769 sq. ft.
Superheater heating surface	1325 sq. ft.
Weight on drivers	304,190 lbs.
Weight on lead truck	42,480 lbs.
Weight on trailing truck	108,640 lbs.
Weight of total engine	455,310 lbs.
Weight of tender	311,700 lbs.
Tender capacity	22 tons; 14,000 gals.
Tractive force, engine	84,600 lbs.
Tractive force, booster	13,300 lbs.
Builder, date	Baldwin, 1930

Collection of H. L. Broadbelt.

Philip R. Hastings.

C&IM KNEW A GOOD THING . . .

Coal-hauling, hill-climbing Chicago & Illinois Midland kept on buying 2-10-2's new or second-hand (from Coast Line and Wabash) long after they'd gone out of style, never regretted the choice for a profile that included a 1.64 per cent grade. Lima-built 701 brightens the Illinois countryside near Kelsey as she heads south for the mines with a mile of empty hoppers behind.

FOR A RED BALL REPUTATION

Wabash was quick to see that a high tractive effort engine which couldn't hold a sustained speed did not enhance its Red Ball service. So in 1930 the road supplanted 64-inch-drivered Mikes with faster but slightly less powerful 4-8-2's and was so pleased that it soon bought 4-8-4's too. An M-1 has time freight Extra 2806 East rolling out of Montpelier, O., for Detroit.

Richard J. Cook.

Collection of George Sennhauser.

CHICAGO & ILLINOIS MIDLAND H-1 SANTA FE (2-10-2) FOR FREIGHT SERVICE

Series	700-701
Cylinders, diameter and stroke	30 x 32 in.
Driving wheel diameter	63 in.
Boiler pressure	200 lbs. per sq. in.
Grate area	88.2 sq. ft.
Evapòrative heating surface	4870 sq. ft.
Superheater heating surface	1285 sq. ft.
Weight on drivers	321,700 lbs.
Weight on lead truck	20,700 lbs.
Weight on trailing truck	63,200 lbs.
Weight of total engine	405,600 lbs.
Weight of tender, loaded	209,300 lbs.
Tender capacity	16 tons; 12,000 gals.
Tractive force, engine	77,700 lbs.
Tractive force, booster	10,500 lbs.
Builder, date	Lima, 1931

WABASH M-1 MOUNTAIN (4-8-2) FOR FREIGHT SERVICE

Series	2800-2824
Cylinders, diameter and stroke	27 x 32 in.
Driving wheel diameter	70 in.
Boiler pressure	235 lbs. per sq. in.
Grate area	84.2 sq. ft.
Evaporative heating surface	4620 sq. ft.
Superheater heating surface	2004 sq. ft.
Weight on drivers	270,400 lbs.
Weight on lead truck	72,100 lbs.
Weight on trailing truck	63,900 lbs.
Weight of total engine	406,400 lbs.
Weight of tender, loaded	291,000 lbs.
Tender capacity	18 tons; 15,000 gals.
Tractive force, engine	66,570 lbs.
Builder, date	Baldwin, 1930

Richard D. Acton.

Dave Gibson.

UNDERSTANDABLY ON ROLLER BEARINGS

Although the 171-mile Akron, Canton & Youngstown doesn't serve Canton, O., home of Timken Roller Bearing, the road was one of several Ohio carriers whose locomotives bore the distinctive Timken journals on engine lead, trailer, and tender trucks. Not long before diesels drop her fires forever AC&Y Lima-built 2-8-2 No. 405 pinch-hits for an FM road-switcher unit at Carey, O.

A LIMA CUSTOMER IN LIMA

Five railroads serve Lima, O., and the Lima Locomotive Works sold steam locomotives to all of them. Following control by Henry Ford, the Detroit, Toledo & Ironton purchased a total of 6 Berkshires and 12 Mikes between 1935 and 1944. Extra 801 South at St. Paris, O., is headed by a Lima 2-8-2 constructed in 1940 which has pilot footboards for switching chores en route.

AKRON, CANTON & YOUNGSTOWN R-2 MIKADO (2-8-2) FOR FREIGHT SERVICE

Number	406
Cylinders, diameter and stroke	26 x 30 in.
Driving wheel diameter	64 in.
Boiler pressure	200 lbs. per sq. in.
Grate area	66.7 sq. ft.
Evaporative heating surface	3507 sq. ft.
Superheater heating surface	1246 sq. ft.
Weight on drivers	233,100 lbs.
Weight on lead truck	36,800 lbs.
Weight on trailing truck	57,400 lbs.
Weight of total engine	327,300 lbs.
Weight of tender, 2/3 load	177,430 lbs.
Tender capacity	16 tons; 12,400 gals.
Tractive force, engine	53,800 lbs.
Builder, date	Lima, 1944

DETROIT, TOLEDO & IRONTON 800 MIKADO (2-8-2) FOR FREIGHT SERVICE

Series	804-807
Cylinders, diameter and stroke	23 x 30 in.
Driving wheel diameter	63 in.
Boiler pressure	260 lbs. per sq. in.
Grate area	66.9 sq. ft.
Evaporative heating surface	4009 sq. ft.
Superheater heating surface	1815 sq. ft.
Weight on drivers	248,500 lbs.
Weight on lead truck	50,100 lbs.
Weight on trailing truck	70,900 lbs.
Weight of total engine	369,500 lbs.
Weight of tender, 2/3 load	194,166 lbs.
Tender capacity	18 tons; 14,300 gals.
Tractive force, engine	55,600 lbs.
Builder, date	Lima, 1941

TRAINS.

MORE BURLY THAN BEAUTIFUL

If Milwaukee Road's 40 S-2 Northerns by Baldwin were weak in esthetics, they were strong in horsepower. On the long hauls out of Bensenville, Ill., to Council Bluffs and to the Twin Cities they worked tonnage with a vengeance and, when the occasion demanded, handled an 18-car *Olympian*. No. 210 whistles for a crossing near Milwaukee on a hot Wisconsin summer day.

FOR THE HEAVIEST OF TRAINS

Few if any steam locomotives have handled heavier train tonnages than Missabe Road's 18 M-3 and M-4 Yellowstones. In the late 1950's they became accustomed to ore runs of more than 19,000 gross tons. In this instance it's a case of the elephant bowing to the gnat as M-3 No. 227 pauses outside Two Harbors, Minn., for a meet with DM&IR's train No. 5, a Budd rail diesel car.

Philip R. Hastings.

CHICAGO, MILWAUKEE, ST. PAUL & PACIFIC S-2 NORTHERN (4-8-4) FOR FREIGHT SERVICE

Series	200-229	Weight on lead truck	87,800 lbs.
Cylinders, diameter and stroke	26 x 32 in.	Weight on trailing truck	120,330 lbs.
Driving wheel diameter	74 in.	Weight of total engine	490,450 lbs.
Boiler pressure	285 lbs. per sq. in.	Weight of tender, loaded	397,000 lbs.
Grate area	106 sq. ft.	Tender capacity	25 tons; 20,000 gals.
Evaporative heating surface	5509 sq. ft.	Tractive force, engine	70,800 lbs.
Superheater heating surface	2336 sq. ft.	Builder, date	Baldwin, 1937
Weight on drivers	282,320 lbs.		

DULUTH, MISSABE & IRON RANGE M-4 YELLOWSTONE (2-8-8-4) FOR FREIGHT SERVICE

Series	228-237	Weight on lead truck	41,350 lbs.
Cylinders, diameter and stroke	26 x 32 in. (4)	Weight on trailing truck	93,350 lbs.
Driving wheel diameter	63 in.	Weight of total engine	699,700 lbs.
Boiler pressure	240 lbs. per sq. in.	Weight of tender, loaded	438,300 lbs.
Grate area	125 sq. ft.	Tender capacity	26 tons; 25,000 gals.
Evaporative heating surface	6758 sq. ft.	Tractive force, engine	140,000 lbs.
Superheater heating surface	2770 sq. ft.	Builder, date	Baldwin, 1943
Weight on drivers	565,000 lbs.		

Jim Scribbins.

Robert Olmsted.

HOW TO PINCH PENNIES

Soo Line, always on a tight budget, spent its dollars wisely in 1938 for just 4 Lima 4-8-4's to cut costs on the 437-mile Chicago-Twin Cities run of subsidiary Wisconsin Central. The O-20's raised gross ton-miles per freight train-hour as much as 30 per cent, replaced 7 Mountains, averaged more than 8300 miles per month per engine. No. 5001 heads north near Vandyne, Wis.

BORN 10 YEARS TOO SOON

On the eve of the 1929 stock market crash that pushed Rock Island into bankruptcy by 1933 the road sampled an Alco 4-8-4, quickly bought a fleet of 65 — the most Northerns anybody ever had in a single class. Excess baggage in a depression, the 5000's were later modernized for laudable war work. Big-tanked 4-8-4 No. 5058 is rolling west near Lawrence, Kans., in 1952.

CHICAGO, ROCK ISLAND & PACIFIC R 67 B NORTHERN (4-8-4) FOR FREIGHT SERVICE

Number	5000	Weight on lead truck	68,000 lbs.
Cylinders, diameter and stroke	26 x 32 in.	Weight on trailing truck	100,500 lbs.
Driving wheel diameter	69 in.	Weight of total engine	434,000 lbs.
Boiler pressure	250 lbs. per sq. in.	Weight of tender	299,200 lbs.
Grate area	88.3 sq. ft.	Tender capacity	20 tons; 15,000 gals.
Evaporative heating surface	5443 sq. ft.	Tractive force, engine	66,700 lbs.
Superheater heating surface	2243 sq. ft.	Tractive force, booster	13,100 lbs.
Weight on drivers	265,500 lbs.	Builder, date	American, 1929

WISCONSIN CENTRAL O-20 NORTHERN (4-8-4) FOR FREIGHT SERVICE

Series	5000-5003	Weight on lead truck	76,000 lbs.
Cylinders, diameter and stroke	26 x 32 in.	Weight on trailing truck	106,500 lbs.
Driving wheel diameter	75 in.	Weight of total engine	445,500 lbs.
Boiler pressure	270 lbs. per sq. in.	Weight of tender, 2/3 load	252,967 lbs.
Grate area	88.3 sq. ft.	Tender capacity	24 tons; 17,500 gals.
Evaporative heating surface	5142 sq. ft.	Tractive force, engine	66,000 lbs.
Superheater heating surface	2120 sq. ft.	Builder, date	Lima, 1938
Weight on drivers	263,000 lbs.		

Collection of George Sennhauser.

Philip R. Hastings.

THE POCAHONTAS REGION

. . . three heavy-duty coal haulers pushed steam power capacity to the screaming point

THE conveyor-belt railroading practiced by that triumvirate of Pocahontas region coal haulers — Chesapeake & Ohio, Norfolk & Western, and Virginian — once produced extraordinary locomotives. Their mission in life was to tote bituminous coal out of the mountain fastness of West Virginia and move it west to the gateways of Deepwater, Columbus, and Toledo and east to tidewater at Hampton Roads. And then conclude the cycle by returning the empty hoppers to the mine tipples. Here, then, was an unexcelled opportunity to hold down the transportation ratio by moving maximum-weight, single-commodity trains — the only qualifications being ruling grades and motive power capacity. Today no more than five diesel units operated in multiple are placed on the head end — that being the practical drawbar load limit — and a Minneapolis & St. Louis can formulate such a locomotive if circumstances warrant. The key, then, to the spectacular engines of the Pocahontas roads in steam days was physical plant. Engine designers could and did specify axle loadings in excess of 80,000 pounds. Consequently, they developed articulateds capable of exerting 7500 drawbar horsepower or a starting tractive effort of 170,000 pounds. And in auxiliary forays into passenger power, they managed the world's heaviest 4-6-4 and the most powerful 4-8-4.

The pores of Chesapeake & Ohio 2-6-6-6 No. 1647 are literally perspiring steam as she starts 90 cars of coal east from engine terminal at Handley, W. Va.

Developmental work flowed in two seldom parallel channels, with the smallest road of the three, Virginian, picking the others' brains and blueprints indiscriminately. Chesapeake & Ohio purchased its power exclusively from commercial builders and relied upon a design development committee which stemmed from Van Sweringen days. Alternating between rigid-frame and articulated power, C&O indulged in different wheel arrangements with great abandon. Norfolk & Western constructed all of its own engines after 1927, never veered from the principle of articulation for road freight service, and concentrated its design efforts on refining just three wheel arrangements. To a greater degree than any other railroad, N&W also improved the operating climate of its locomotives with modern servicing facilities and utilization techniques.

Famous theretofore mainly for introducing the 4-8-2 to the U. S. in 1911, Chesapeake & Ohio distinguished itself in 1924 by beginning to buy the first fleet of single-expansion articulateds in the land. Its 50 H-7 2-8-8-2's (30 of which found their way onto the Union Pacific roster in 1945) were simple mainly because a tunnel in the Alleghenies would not accommodate the low-pressure cylinders of any Mallet larger than a 2-6-6-2. Impressive as the long-boilered engines were, though, they were eclipsed within six years by the Lima Super-Power T-1 2-10-4. C&O's Texas, with booster, managed to equal the tractive force of the 2-8-8-2 despite a jump in driver size (hence train speed) from 57 inches to 69. The T-1's worked mainly between Russell, Ky., and Toledo, O. Five 4-8-4's for mountain passenger service came from Lima in 1935, but otherwise Chessie didn't purr in the design department until 1941. What came then was a landmark in steam locomotive development.

Chesapeake & Ohio required stauncher power for its short but stiff 80-mile Allegheny Subdivision between Hinton, W. Va., and Clifton Forge, Va., where the ruling grade against coal loads on the west slope was .57 per cent (1.14 per cent on the east slope) and there were 6-degree curves. The simple solution would have been a heavier, slightly longer version of the lumbering H-7. Instead C&O ushered in a totally new wheel arrangement — the 2-6-6-6 — which was not only Lima's largest product but the heaviest, most powerful six-coupled articulated ever built. In service, two of the new H-8's (appropriately labeled Alleghenies) could lift 11,500 tons up the .57 per cent grade to the 2072-foot summit of their namesake mountains. Out in Ohio on test on level, tangent track a single 2-6-6-6 once moved a 160-car, 14,083-ton coal train to 19 miles per hour from a standing start in 1 mile and 6 minutes, and had it rolling 29 miles per hour within a total elapsed time of 11 minutes. Between 1941 and 1948 Lima constructed 60 H-8's (plus 8 almost identical but even heavier 2-6-6-6's for Virginian), and it is entirely fitting that the prototype, No. 1601, is suitably preserved under cover at the Henry Ford Museum in Dearborn, Mich.

During the war and right on up through 1948 C&O kept buying other "firsts" from all three builders. Its ponderous Baldwin Hudsons, the world's heaviest of course, had an engine weight only 7500 pounds less than that of a Canadian Pacific 2-10-4! Although C&O's Kanawhas were based upon the 2-8-4's of Van Sweringen associates Nickel Plate and Pere Marquette, they were the heaviest of their wheel arrangement, too, and 90 of them were delivered in only four years, not to mention an order for five more from Virginian. And there were additional 4-8-4's, 0-8-0's, and even mine-run compound 2-6-6-2's being added to the roster clear up until May 1949, when C&O reluctantly bought a few diesel yard units. A year later the road decided to go whole hog, and the death knell was sounded for the last grand stand of standard railroading.

Norfolk & Western was a workhorse of another strain. It dieselized in theory if not in practice a couple of decades before the first Geep hit the property. Although passionately dedicated to the coal-burning locomotive, the railway was not unaware of the threat to steam's existence (and incidentally to such coal as was carried for off-line locomotive consumption). So in 1936 N&W began building in its Roanoke (Va.) Shops what it termed "modern" locomotives (*i.e.*, those with big boilers, one-piece cast steel bed frames, roller bearings, improved counterbalancing, and mechanical and pressure lubrication). On N&W they came in just three convenient sizes: the streamlined 70-inch-drivered J-class 4-8-4 for passenger work; the dual-purpose, single-expansion A-class 2-6-6-4, also fitted with a 70-inch wheel; and most numerous, the gutty, 57-inch-drivered Y6 compound 2-8-8-2. Each of the three was, in its own right, a remarkable design. Together they held the dike against dieselization the longest with the least leakage.

N&W knew when to pioneer . . . and when not to. Its 2-6-6-4 was one of the first satisfactory high-speed simple articulateds as well as one of the finest, an engine that could average 30 miles per hour start to stop with 14,500 tons of solid coal, yet roll either Pullmans or 125 cars of merchandise at twice that speed. Its dreadnaught compound 2-8-8-2, on the other hand, could be traced back in essential design to World War I and the N&W Y2 picked as the standardized U.S.R.A. Mallet. Instead of discarding the principle of using steam twice because of its low-speed limitations, N&W capitalized on compounding, finally achieved a machine of 170,000 pounds initial tractive force that could be worked in simple gear up to 10 miles per hour and even given a shot of high-pressure steam in her l.p. cylinders when the going got rough on the grades. In retrospect, more than one mountain operator of 4-6-6-4's secretly must have wished to reconsider. As for the J, need anything more be said than that a comparatively low-drivered 4-8-4 proved herself not only equal to the mountains but also capable of whipping a 1025-ton test passenger train up to 110 miles per hour across the Virginia swamps?

O. Winston Link.

Such locomotives were exploited to the hilt by engine terminals which could service a Mallet in less than an hour; by the widespread adoption of auxiliary water tanks; and by an operating department that dropped the fire of any engine which stood idle too long. Accordingly, the Y6's could average 6000 miles a month in drag service; and on a railway whose longest engine run was the 676½ miles between Cincinnati and Norfolk the 4-8-4's managed 15,000 miles a month. In 1953 N&W's 143 modern articulateds accounted for 94 per cent of its gross ton-miles, and its 14 J's ran off 84 per cent of passenger train-miles.

Asked once if his works had ever attempted to sell a new boiler concept to the N&W, a Lima man shrugged and said, "What could we tell them about steam power?" Such was the respect Roanoke won for itself.

Of such material — Super-Power in full stride and steam locomotion treated as a science — was the memorable Pocahontas region record created. I

Philip A. Weibler.

LIMA'S FINEST HOUR

The builder which coined the term "Super-Power" poured all of its genius into Chesapeake & Ohio's 2-6-6-6, a brand-new wheel arrangement supporting a boiler of phenomenal capacity. In 1956 the camera finds 1624 on an extra near Thurmond, W. Va., along the New River, after diesels had bumped her off the Alleghenies climb.

HEAVIEST HUDSONS EVER BUILT

Chesapeake & Ohio's enormous L-2 4-6-4's with Franklin rotary-cam poppet valves boasted an engine weight of 221½ tons and were coal-minded Chessie's last bid to beat the passenger diesel with a reciprocating engine. The 314 grinds to a halt at Russell, Ky., with the *F.F.V.* She was the newest, biggest, and last 4-6-4.

W. H. N. Rossiter.

Collection of M. E. Kern.

CHESAPEAKE & OHIO L-2 HUDSON (4-6-4) FOR PASSENGER SERVICE

Series	310-314	Weight on lead truck	94,500 lbs.
Cylinders, diameter and stroke	25 x 30 in.	Weight on trailing truck	129,000 lbs.
Driving wheel diameter	78 in.	Weight of total engine	443,000 lbs.
Boiler pressure	255 lbs. per sq. in.	Weight of tender, 2/3 load	317,500 lbs.
Grate area	90.2 sq. ft.	Tender capacity	30 tons; 21,000 gals.
Evaporative heating surface	4178 sq. ft.	Tractive force, engine	52,100 lbs.
Superheater heating surface	1785 sq. ft.	Tractive force, booster	14,200 lbs.
Weight on drivers	219,500 lbs.	Builder, date	Baldwin, 1948

CHESAPEAKE & OHIO H-8 ALLEGHENY (2-6-6-6) FOR FREIGHT SERVICE

Series	1600-1609	Weight on lead truck	64,500 lbs.
Cylinders, diameter and stroke	22½ x 33 in. (4)	Weight on trailing truck	189,000 lbs.
Driving wheel diameter	67 in.	Weight of total engine	724,500 lbs.
Boiler pressure	260 lbs. per sq. in.	Weight of tender, 2/3 load	341,600 lbs.
Grate area	135.2 sq. ft.	Tender capacity	25 tons; 25,000 gals.
Evaporative heating surface	7240 sq. ft.	Tractive force, engine	110,200 lbs.
Superheater heating surface	3186 sq. ft.	Builder, date	Lima, 1941
Weight on drivers	471,000 lbs.		

David W. Salter.

COPIES OF THE NEIGHBOR'S ENGINE

As the youngster of the Pocahontas threesome of coal haulers, Virginian was usually content simply to duplicate the big steam power of its neighbors. Lima 2-8-4 508, fighting hard here to keep westbound empties rolling on a grade at Suffolk, Va., was one of five virtual carbon copies of Chesapeake & Ohio's big K-4 Kanawha.

AMONG THE FINEST OF 2-8-4'S

Swinging around a curve with solid coal tied to her tank, Kanawha 2705 of Chesapeake & Ohio typifies a fleet of 90 2-8-4's of extraordinary size and flexibility. C&O employed the roller-bearing machines on mineral trains, merchandisers, and passenger schedules — never found the K-4 class short on steam or speed or punch.

VIRGINIAN BA BERKSHIRE (2-8-4) FOR FREIGHT SERVICE

Series	505-509
Cylinders, diameter and stroke	26 x 34 in.
Driving wheel diameter	69 in.
Boiler pressure	245 lbs. per sq. in.
Grate area	90.3 sq. ft.
Evaporative heating surface	4774 sq. ft.
Superheater heating surface	1932 sq. ft.
Weight on drivers	295,600 lbs.
Weight on lead truck	46,000 lbs.
Weight on trailing truck	118,800 lbs.
Weight of total engine	460,400 lbs.
Weight of tender, 2/3 load	326,800 lbs.
Tender capacity	21 tons; 25,000 gals.
Tractive force, engine	69,350 lbs.
Builder, date	Lima, 1946

CHESAPEAKE & OHIO K-4 KANAWHA (2-8-4) FOR FREIGHT SERVICE

Series	2760-2789
Cylinders, diameter and stroke	26 x 34 in.
Driving wheel diameter	69 in.
Boiler pressure	245 lbs. per sq. in.
Grate area	90.3 sq. ft.
Evaporative heating surface	4771 sq. ft.
Superheater heating surface	1940 sq. ft.
Weight on drivers	292,500 lbs.
Weight on lead truck	46,200 lbs.
Weight on trailing truck	124,800 lbs.
Weight of total engine	463,500 lbs.
Weight of tender, 2/3 load	308,100 lbs.
Tender capacity	30 tons; 21,000 gals.
Tractive force, engine	69,350 lbs.
Tractive force, booster	14,000 lbs.
Builder, date	American, 1947

Collection of M. E. Kern.

Tom Miller.

William E. Warden Jr.

COMPOUNDING . . . BROUGHT UP TO DATE

From World War I the backbone of the Norfolk & Western roster was a compound 2-8-8-2 with a long and illustrious history. Sired by the N&W-designed Y2 (selected as the standardized U.S.R.A. Mallet), the latter-day Y6 demonstrated what could be done by sticking with a good thing. Homemade 2179 shoves hard on the tail of a freight climbing Blue Ridge's 1.2 per cent.

SINGLE EXPANSION, DOUBLE DUTY

Few articulated locomotives proved themselves so versatile as Norfolk & Western's sharp-voiced A-class simple 2-6-6-4. An A engine could roll a passenger train comfortably at 70 miles per hour, or iron out the hills with 125 cars of manifest freight, or handle 14,500 tons of coal west from Williamson, W. Va. No. 1224 is keeping N&W's shippers happy with a time freight.

Collection of M. E. Kern.

NORFOLK & WESTERN A 2-6-6-4 FOR FREIGHT SERVICE

Series	1238-1242
Cylinders, diameter and stroke	24 x 30 in. (4)
Driving wheel diameter	70 in.
Boiler pressure	300 lbs. per sq. in.
Grate area	122 sq. ft.
Evaporative heating surface	6639 sq. ft.
Superheater heating surface	2703 sq. ft.
Weight on drivers	432,350 lbs.
Weight on lead truck	30,480 lbs.
Weight on trailing truck	110,170 lbs.
Weight of total engine	573,000 lbs.
Weight of tender, loaded	378,600 lbs.
Tender capacity	30 tons; 22,000 gals.
Tractive force, engine	114,000 lbs.
Builder, date	Roanoke, 1949-1950

NORFOLK & WESTERN Y6b 2-8-8-2 FOR FREIGHT SERVICE

Series	2171-2200
Cylinders, dia. and stroke	25 and 39 x 32 in. (4)
Driving wheel diameter	58 in.
Boiler pressure	300 lbs. per sq. in.
Grate area	106.2 sq. ft.
Evaporative heating surface	4915 sq. ft.
Superheater heating surface	1478 sq. ft.
Weight on drivers	522,850 lbs.
Weight on lead truck	32,200 lbs.
Weight on trailing truck	27,850 lbs.
Weight of total engine	582,900 lbs.
Weight of tender, loaded	378,600 lbs.
Tender capacity	30 tons; 22,000 gals.
Tractive force, engine	152,206 lbs. simple 126,838 lbs. compound
Builder, date	Roanoke, 1948-1952

Rail Photo Service: G. C. Corey.

WORLD'S MOST POWERFUL 4-8-4

Norfolk & Western J-class 4-8-4 612 works her way west through West Virginia with train No. 9, a Bluefield-Williamson local. Once the pressure of the comparatively low-drivered J's was raised to 300 pounds, they ranked as the brawniest of 4-8-4's with a tractive force of 80,000 pounds. Streamlining concealed girth temporarily visible in a few stripped models built in the war.

Bill Taub.

AS IT WOULD HAVE BEEN MINUS DIESELS

Until Chesapeake & Ohio's sudden change of corporate heart in behalf of diesels, *Train X*, and blue and yellow paint, it operated in the grand old tradition, buying ever larger steam engines and 12-wheel Pullman-green coaches for them to pull. Lima Greenbrier 610 swings her roller-bearing rods over Hampton Creek with the westbound *George Washington*'s Virginia section.

M. E. Kern.

CHESAPEAKE & OHIO J-3-A GREENBRIER (4-8-4) FOR PASSENGER SERVICE

Series	610-614	Weight on lead truck	81,600 lbs.
Cylinders, diameter and stroke	27½ x 30 in.	Weight on trailing truck	115,400 lbs.
Driving wheel diameter	72 in.	Weight of total engine	479,400 lbs.
Boiler pressure	255 lbs. per sq. in.	Weight of tender, 2/3 load	309,700 lbs.
Grate area	100.3 sq. ft.	Tender capacity	25 tons; 21,500 gals.
Evaporative heating surface	4823 sq. ft.	Tractive force, engine	68,300 lbs.
Superheater heating surface	2058 sq. ft.	Tractive force, booster	12,400 lbs.
Weight on drivers	282,400 lbs.	Builder, date	Lima, 1948

NORFOLK & WESTERN J NORTHERN (4-8-4) FOR PASSENGER SERVICE

Series	600-604	Weight on lead truck	90,000 lbs.
Cylinders, diameter and stroke	27 x 32 in.	Weight on trailing truck	116,000 lbs.
Driving wheel diameter	70 in.	Weight of total engine	494,000 lbs.
Boiler pressure	275 lbs. per sq. in.	Weight of tender, loaded	378,600 lbs.
Grate area	107.7 sq. ft.	Tender capacity	26 tons; 22,000 gals.
Evaporative heating surface	5271 sq. ft.	Tractive force, engine	73,300 lbs.
Superheater heating surface	2177 sq. ft.	Tractive force, booster	12,500 lbs.
Weight on drivers	288,000 lbs.	Builder, date	Roanoke, 1941-1942

John Krave.

THE SOUTH

. . . comparatively few new engines were purchased after 1930 — but these are remembered

THE South was content with a rear-guard position in locomotive development until approximately World War II, and then the diesel won acceptance for reasons that had balked application of modern steam power. Predominantly an agricultural economy until the war, the South had seldom enjoyed the traffic density which elsewhere had underwritten big engines (and physical plants to accommodate them). Yet just when the South was shifting its focus from fields to factories, the multiple-unit diesel made available high horsepower without restrictive axle loadings. No wonder, then, that Gulf, Mobile & Ohio, whose biggest engine had been a Decapod, became the first sizable class 1 railroad to achieve total dieselization. And it was just as logical that the original diesel freight demonstrator should wind up on the Southern, which had been previously unable to operate anything larger than a 2-8-0 on its main lines into St. Louis and Jacksonville.

In a sense, the final epoch of steam locomotive development in the South began in 1918 when the region accepted the United States Railroad Administration's standardized designs with ill-concealed enthusiasm. A Pennsy or a Milwaukee Road might have quibbled with the "McAdoos," but south of the Ohio River the Government engines raised rather than conflicted with the level of motive power design.

Coal from mines on Louisville & Nashville's Eastern Kentucky Division is hauled through a rustic setting at Wallins, Ky., by roller-bearing Berkshire 1986.

Accordingly, the region continued to purchase the standardized locomotives right up until the bubble burst in 1929-1930, pausing only to alter a driving wheel diameter here and perhaps add a Delta trailing truck there. Louisville & Nashville eventually amassed 234 engines of U.S.R.A. pattern in five wheel arrangements from 0-8-0 to 4-8-2. Atlantic Coast Line bought 235 duplicates of a single prototype, the U.S.R.A. light 4-6-2, and coupled them to passengers and perishables with equal aplomb. Southern swarmed with unmistakably U.S.R.A. power, although many of the big Mikes and Pacifics tried to hide their origin behind apple green Duco or Elesco feedwater heaters.

There were exceptions. Nashville, Chattanooga & St. Louis did buy five small 4-8-4's (dubbed Dixies, never Northerns) from Alco in 1930, and the Chicago-headquartered Illinois Central was so thunderstruck by a test of the Lima A-1 2-8-4 that it bought 50 duplicate engines plus the demonstrator herself.

After 1930 most railroads in the South turned a deaf ear to the builders until diesels came. Fewer than 180 new steam locomotives were acquired after the Wall Street collapse. The 8000-mile Southern, for example, never had an engine with a four-wheel trailing truck on the property and bought its last new steam engine in 1928. Between them, Seaboard and Coast Line purchased only 22 modern steam locomotives. Marginal properties such as the Georgia and the Tennessee Central also made do with what they had or, like the Atlanta, Birmingham & Coast, bought power secondhand.

Of course, if the diesel hadn't come . . . but it did.

It would be an injustice, however, to dismiss the South on the basis of these statistics. The record indicates, for example, that Illinois Central bought only 20 new freight engines between its Limas of 1926 and the first GP7 delivered in 1950; these were the homemade 2600-series 4-8-2's built by Paducah (Ky.) Shops in 1942-1943. The explanation is that instead of buying new engines the IC embarked upon the largest rebuilding program in modern times. A class of 125 2-10-2 drag engines purchased in the early 1920's were either (1) souped up by increased boiler pressure into "Central" types of the same wheel arrangement or (2) turned into higher-drivered 4-8-2's by the application of new one-piece cast steel frames. Mikados were modernized in bewildering ways, sometimes emerging from Paducah as 2-10-0's or 0-8-2's for heavy-duty terminal work, and in other programs remaining 2-8-2's despite application of former 2-10-2 cylinders and frames for increased tractive force. Indeed, Paducah became so adept at such major surgery that one suspects that the shops must have regarded a simple birth such as that of the new 2600's as routine. Certainly Paducah should receive sole credit for the extraordinary accomplishment of having added more than 18 million pounds of tractive force to an existing roster of locomotives.

Competitor Louisville & Nashville, equally coal-minded until it had no choice but diesels, got by with nothing heavier than a 2-8-2 until 1942. The selection then was a 2-8-4 (partially because South Louisville Shops' erection floor wasn't wide enough for a 4-8-4) — an engine quite as notable for its modernity as its size. The M-1's had roller bearings on all axles, high-speed Franklin boosters, steam lines, tender coal pushers — all the gadgets one could ask. Among the heaviest Berkshires ever built, they were reputed to be the most expensive too. L&N operated them in eastern Kentucky in an extremely wide range of service which ran the gamut from passenger schedules to coal trains to helper assignments. The sight and sound of three M-1's hoisting 8200 tons upgrade from the Kentucky River at Ford, Ky., was an experience, for few 69-inch-drivered locomotives lugged so well.

The dual-purpose 4-8-4 was an obvious choice for the South's bridge roads: Nashville, Chattanooga & St. Louis; and Richmond, Fredericksburg & Potomac. Both had previously employed several wheel arrangements to forward an "overhead" traffic of both freight and passengers; the 4-8-4 could alternate between both jobs and reduce doubleheading to a minimum. NC&StL augmented its original 4-8-4's of 1930 with 20 modernized, roller-bearing engines of similar specifications in 1942-1943 — some of them dubbed "Yellow Jackets" because of their skirted, painted running boards. Racing over a route made immortal by the *General* and the *Texas*, these J-3's kept a vital line fluid during the crisis years of world war. RF&P's 27 4-8-4's, all purchased from Baldwin between 1937 and 1945, were so formidably handsome that they appeared heavier than they were. Named for governors, Confederate generals, and other prominent Virginians, the 550's and 600's shuttled like streetcars over a 113.5-mile main. In point of fact, both the NC&StL and the RF&P machines were relatively lightweight 4-8-4's, more equivalent to Canadian power than to most domestic Northerns. Nevertheless, they worked wonders with solid strings of standard sleepers. In fast passenger service a Governor was rated at 2200 tons southbound and 2900 tons northbound between Washington Union Station and Richmond Broad Street.

Coast Line must have been impressed by RF&P's Baldwins. In 1938 ACL purchased a dozen remarkably similar 4-8-4's. Prototype No. 1800 quickly caused a flurry of news when she sliced anticipated acceleration times in half with a 20-car, 1500-ton *Havana Special* and reeled off 5 miles at an average speed of 84 miles per hour. But the 1800's proved rough on the track and less of a tourist attraction than rival Seaboard's diesels; they were only a year old when the first road diesels came to the property. Seaboard itself pioneered in steam by buying 10 69-inch-drivered 2-6-6-4's, which were the first fleet of articulateds specifically designed for high-speed freight and optional passenger service. They finished their days on B&O following SAL's dieselization. A broader application of modern articulated power in the South occurred on Clinchfield, where 4-6-6-4's of D&H and UP ancestry rubbed shoulders.

Bruce Meyer.

Nor should we forget those vest-pocket Berkshires of Norfolk Southern (which wound up in Mexico) . . . Central of Georgia's tiny-tanked 4-8-4's (which were really Espee GS-2's minus the streamlining) . . . a pair of West Point Mikes (one of which held down the *Crescent* for a spell) . . . and 10 Lima 2-8-4's for RF&P (which turned its back suddenly on the dual-purpose idea).

If the South was quantitatively in the rear guard of steam power development but a quick convert to dieselization, at least the region had its renegades. Water-level grades, Paducah's rebuilding program, and a shrewd bid for competitive coal traffic prolonged Illinois Central's D-Day for years . . . Lima was building 2-8-4's for Louisville & Nashville as late as 1949. The invincible diesel finally conquered all, but it was hardly a second march to the sea. I

August A. Thieme Jr.

J. P. Lamb Jr.

INTENDED FOR THE SUN WORSHIPERS

Handsome, obviously Baldwin-built 1800-1811 series of Coast Line 4-8-4's were purchased to supplement aging Pacifics on Florida limiteds in 1938. Their tendency to pound the rail, plus ACL's acquisition of diesels, soon bumped the R-1's into freight service. Northern 1808 has an extra rolling fast south of Richmond in 1947.

SQUARE SANDBOX, DISC DRIVERS

Illinois Central's Paducah (Ky.) Shops became so experienced at major locomotive rebuilding (*e.g.*, turning 2-10-2's into 4-8-2's) that few were surprised in 1942-1943 when the road completed 20 brand-new Mountain types. Quite similar to the 2500-series 4-8-2's except for their 12-wheel tanks, the big jobs possess high tractive force. No. 2613 nears Mounds, Ill., on tonnage in 1959.

ILLINOIS CENTRAL MOUNTAIN (4-8-2) FOR FREIGHT SERVICE

Series	2600-2619
Cylinders, diameter and stroke	28 x 30 in.
Driving wheel diameter	70 in.
Boiler pressure	275 lbs. per sq. in.
Grate area	88.3 sq. ft.
Evaporative heating surface	5194.7 sq. ft.
Superheater heating surface	1242 sq. ft.
Weight on drivers	293,880 lbs.
Weight on lead truck	61,193 lbs.
Weight on trailing truck	68,820 lbs.
Weight of total engine	423,893 lbs.
Weight of tender	370,300 lbs.
Tender capacity	26 tons; 22,000 gals.
Tractive force, engine	83,160 lbs.
Builder, date	Paducah, 1942-1943

ATLANTIC COAST LINE R-1 NORTHERN (4-8-4) FOR PASSENGER SERVICE

Series	1800-1811
Cylinders, diameter and stroke	27 x 30 in.
Driving wheel diameter	80 in.
Boiler pressure	275 lbs. per sq. in.
Grate area	97.75 sq. ft.
Evaporative heating surface	4749 sq. ft.
Superheater heating surface	1497 sq. ft.
Weight on drivers	263,127 lbs.
Weight on lead truck	89,343 lbs.
Weight on trailing truck	107,800 lbs.
Weight of total engine	460,270 lbs.
Weight of tender, loaded	435,500 lbs.
Tender capacity	27 tons; 24,000 gals.
Tractive force, engine	63,900 lbs.
Builder, date	Baldwin, 1938

Collection of H. L. Broadbelt.

John A. Krave.

A VERY VERSATILE BERKSHIRE

Chances are that the tonnage exceeds 8000 as Louisville & Nashville 2-8-4 1989 gets a coal train under way west from Cumberland, Ky., on the Cumberland Valley Division. The 69-inch wheel of the M-1 proved quite satisfactory for passenger service and the 2-8-4 made a homogeneous helper for a train powered by a sister.

Rail Photo Service: Shelby Lowe.

A DIXIE ON THE DIXIE LIMITED

Flange-stacked and cone-nosed, a Nashville, Chattanooga & St. Louis J-3 4-8-4 gives the mail clerks a look at a trim Alco as she leads the *Dixie Limited* out of Smyrna, Ga. The Dixie Line introduced the 4-8-4 wheel arrangement to the South in 1930, found that its dual-purpose nature simplified the task of engine assignment.

LOUISVILLE & NASHVILLE M-1 BERKSHIRE (2-8-4) FOR FREIGHT SERVICE

Series	1970-1991	Weight on lead truck	52,900 lbs.
Cylinders, diameter and stroke	25 x 32 in.	Weight on trailing truck	127,700 lbs.
Driving wheel diameter	69 in.	Weight of total engine	448,100 lbs.
Boiler pressure	265 lbs. per sq. in.	Weight of tender, 2/3 load	308,900 lbs.
Grate area	90.2 sq. ft.	Tender capacity	25 tons; 22,000 gals.
Evaporative heating surface	4653 sq. ft.	Tractive force, engine	65,290 lbs.
Superheater heating surface	1908 sq. ft.	Tractive force, booster	14,000 lbs.
Weight on drivers	267,500 lbs.	Builder, date	Lima, 1949

NASHVILLE, CHATTANOOGA & ST. LOUIS J3-57 DIXIE (4-8-4) FOR DUAL SERVICE

Series	580-589	Weight on lead truck	75,500 lbs.
Cylinders, diameter and stroke	25 x 30 in.	Weight on trailing truck	95,500 lbs.
Driving wheel diameter	70 in.	Weight of total engine	399,000 lbs.
Boiler pressure	250 lbs. per sq. in.	Weight of tender, 2/3 load	234,700 lbs.
Grate area	77.3 sq. ft.	Tender capacity	16 tons; 15,000 gals.
Evaporative heating surface	4203 sq. ft.	Tractive force, engine	57,000 lbs.
Superheater heating surface	1721 sq. ft.	Builder, date	American, 1943
Weight on drivers	228,000 lbs.		

Floyd A. Bruner.

D&H DESIGN DEEP IN DIXIE

For one cause or another the Clinchfield was usually obliged to borrow blueprints for its heavy power. W.P.B. restrictions prevented an original design when wartime traffic necessitated new engines to spell U.S.R.A. 2-8-8-2's, so the bridge route accepted a beefed-up version of Delaware & Hudson's J-95 4-6-6-4. Extra 660 North's Challenger takes siding for a hotshot at Marion, N. C.

NEW CONCEPT FOR ARTICULATEDS

Smokebox-mounted pumps made Seaboard's 2-6-6-4's look like tonnage maulers, perhaps, but the R-1's were bought to roll hotshots and handle passenger trains when required — a new concept for an articulated in 1935. Between Richmond, Va., and Raleigh, N. C., 254 miles, the 2500's bettered the tonnage ratings of 4-8-2's 50 per cent. The 2502 lifts her pops in Raleigh.

Homer R. Hill.

Collection of H. L. Broadbelt.

SEABOARD AIR LINE R-1 2-6-6-4 FOR FREIGHT SERVICE

Series	2500-2504
Cylinders, diameter and stroke	22 x 30 in. (4)
Driving wheel diameter	69 in.
Boiler pressure	230 lbs. per sq. in.
Grate area	96.3 sq. ft.
Evaporative heating surface	5513 sq. ft.
Superheater heating surface	2398 sq. ft.
Weight on drivers	330,000 lbs.
Weight on lead truck	40,000 lbs.
Weight on trailing truck	110,000 lbs.
Weight of total engine	480,000 lbs.
Weight of tender	301,300 lbs.
Tender capacity	24 tons; 16,000 gals.
Tractive force, engine	82,300 lbs.
Builder, date	Baldwin, 1935

CLINCHFIELD E-1 CHALLENGER (4-6-6-4) FOR FREIGHT SERVICE

Series	650-657
Cylinders, diameter and stroke	22 x 32 in. (4)
Driving wheel diameter	69 in.
Boiler pressure	265 lbs. per sq. in.
Grate area	108 sq. ft.
Evaporative heating surface	5392 sq. ft.
Superheater heating surface	1680 sq. ft.
Weight on drivers	420,000 lbs.
Weight on lead truck	79,000 lbs.
Weight on trailing truck	108,000 lbs.
Weight of total engine	607,000 lbs.
Weight of tender, 2/3 load	316,900 lbs.
Tender capacity	26 tons; 22,500 gals.
Tractive force, engine	101,120 lbs.
Builder, date	American, 1942-1943.

A BIG APPLE MOVES THE MELONS

Into the dawn of a hot summer day comes Central of Georgia 4-8-4 No. 452 with a string of ventilated box cars loaded with watermelons. The site: Lorane, Ga. Central's Lima-built K engines (known affectionately as "Big Apples" by CofG employees) were essentially a coal-burning version of Southern Pacific's GS-2 — minus streamlining. Central bought 8 of them.

Hugh M. Comer.

C. W. Witbeck.

A GOVERNOR BOUND FOR THE CAPITAL

Richmond, Fredericksburg & Potomac revived an honored custom when it named as well as numbered all of its handsome Baldwin 4-8-4's. Names of generals, governors, and other prominent Virginians were painted on the cab sides of locomotives that handled passengers and tonnage with equal facility. No. 602, *Governor Thomas Jefferson*, is making time with an 18-car limited.

RICHMOND, FREDERICKSBURG & POTOMAC 4-8-4 FOR PASSENGER SERVICE

Series	607-612	Weight on lead truck	63,100 lbs.
Cylinders, diameter and stroke	27 x 30 in.	Weight on trailing truck	82,500 lbs.
Driving wheel diameter	77 in.	Weight of total engine	408,400 lbs.
Boiler pressure	260 lbs. per sq. in.	Weight of tender, 2/3 load	252,000 lbs.
Grate area	86.5 sq. ft.	Tender capacity	17 tons; 18,000 gals.
Evaporative heating surface	4294 sq. ft.	Tractive force, engine	62,800 lbs.
Superheater heating surface	1325 sq. ft.	Builder, date	Baldwin, 1942
Weight on drivers	262,800 lbs.		

CENTRAL OF GEORGIA K NORTHERN (4-8-4) FOR DUAL SERVICE

Series	451-458	Weight on lead truck	78,900 lbs.
Cylinders, diameter and stroke	27 x 30 in.	Weight on trailing truck	108,300 lbs.
Driving wheel diameter	73½ in.	Weight of total engine	447,200 lbs.
Boiler pressure	250 lbs. per sq. in.	Weight of tender, 2/3 load	196,500 lbs.
Grate area	90.2 sq. ft.	Tender capacity	21 tons; 13,000 gals.
Evaporative heating surface	4705 sq. ft.	Tractive force, engine	63,200 lbs.
Superheater heating surface	2059 sq. ft.	Builder, date	Lima, 1943
Weight on drivers	260,000 lbs.		

Bruce Meyer.

IN THE BEGINNING, THE 2-10-2

The 125 2-10-2's Illinois Central purchased in the 1920's prompted all manner of rebuildings years later in Paducah (Ky.) Shops. Their boilers were used for 4-8-2's; cylinders and frames were tagged for big Mikes; and many were rebuilt into super 2-10-2's called the Central type. Such an engine, the 2808, is rolling west through Princeton, Ky., with empty hoppers behind her.

U.S.R.A. ALL OVER AGAIN

When war pinched the motive power of the West Point Route (Atlanta & West Point; Western Railway of Alabama), each partner road bought a 2-8-2 from Baldwin in 1944. The Mikes were precisely what they looked like: updated heavy U.S.R.A. 2-8-2's. WofA No. 380 is westbound through Red Oak, Ga., in October 1948 with the tonnage of First 211 behind her 12-wheel tender.

Hugh M. Comer.

ATLANTA & WEST POINT AND WESTERN RAILWAY OF ALABAMA F-63 MIKADO (2-8-2) FOR FREIGHT SERVICE

Series	430, 380	Weight on lead truck	28,500 lbs.
Cylinders, diameter and stroke	27 x 32 in.	Weight on trailing truck	48,500 lbs.
Driving wheel diameter	63 in.	Weight of total engine	337,000 lbs.
Boiler pressure	200 lbs. per sq. in.	Weight of tender, 2/3 load	273,000 lbs.
Grate area	70.3 sq. ft.	Tender capacity	18 tons; 18,000 gals.
Evaporative heating surface	4000 sq. ft.	Tractive force, engine	63,000 lbs.
Superheater heating surface	1276 sq. ft.	Builder, date	Baldwin, 1944
Weight on drivers	260,000 lbs.		

ILLINOIS CENTRAL 2850 CENTRAL (2-10-2) FOR FREIGHT SERVICE

Number	2850	Weight on trailing truck	64,360 lbs.
Cylinders, diameter and stroke	30 x 32 in.	Weight of total engine	420,389 lbs.
Driving wheel diameter	64½ in.	Weight of tender	215,500 lbs.
Boiler pressure	260 lbs. per sq. in.	Tender capacity	19 tons; 12,000 gals.
Grate area	88.3 sq. ft.	Tractive force, engine	104,484 lbs.
Evaporative heating surface	5225.5 sq. ft.	Tractive force, booster	13,540 lbs.
Superheater heating surface	1285 sq. ft.	Builder, date	Lima, 1920; rebuilt Paducah, 1943
Weight on drivers	332,500 lbs.		
Weight on lead truck	23,529 lbs.		

R. S. Plummer.

THE SOUTHWEST

. . . birthplace of the Texas type and a hotbed of homemade 4-8-2's and 4-8-4's

DURING the depression many a railroad had much in common with the vagrant who knows that the indignity of a jail sentence carries with it the compensations of a dry cell and three square meals a day. Once a road became insolvent the bankruptcy courts served to keep the creditors at arm's length so that the trustees could spend any surplus after expenses for property improvements. The experience of the Southwest illustrates the point. Moreover, one needs no *Moody's* to tell the players; their motive power rosters explain all. Katy, which barely survived the depression (and finally all but fell apart a few years after the war), had purchased its last new steam engines—10 Lima 0-8-0's — in 1925, thereafter made it all the way to dieselization without resort to the builders. Stronger Kansas City Southern bought just 10 engines, Lima 2-10-4's, in 1937, otherwise made do with its old, chunky, drag-era 2-8-0's and Mallets. Texas & Pacific had salted the profits from an oil boom into a fleet of 95 new locomotives during 1925-1929 — and this power sustained the road nicely until diesels began arriving in 1947.

So much for the solvents. The bankrupts constituted an iron horse of another color. One after another, Missouri Pacific, Frisco, and Cotton Belt fell into the courts. Not long thereafter interesting work was afoot in their respective shops at Sedalia and Springfield, Mo., and Pine Bluff, Ark.—very interesting, indeed.

Baldwin 4-8-4 No. 809 is talking it up through Sulphur Springs, Tex., as she rolls west into the sun with Cotton Belt's Blue Streak *hotshot, First 343.*

Mopac cut its teeth on locomotive improvement by refurbishing 4-4-2's, 4-6-2's, and 2-8-2's with such betterments as disc center drivers, feedwater heaters, solid pilots, and boosters. Next the road upgraded seven light U.S.R.A. Mountains into high-capacity roller-bearing machines on 75-inch drivers. Sedalia Shops excelled itself in 1940 when it began rebuilding a class of 10-year-old Lima 2-8-4's into 4-8-4's. Excluding the boiler shell (which was extended) and such components as stoker, power reverse, air pumps, dome, and sand box, the 2100-series Northerns were virtually new power. New cast-steel engine beds were machined to accommodate 75-inch disc drivers with roller bearings; fireboxes were replaced and tubes rearranged. The dual-purpose 2101-2125 series constituted as dramatic a step forward in steam design as its 2-8-4 germ had a decade previous. On the St. Louis-Pueblo (Colo.) run in freight service the rebuilds began piling up more than 8000 miles per month per engine, a figure which surpassed the Berks' performance by a ratio of 2.34 to 1.

Neighbor Frisco proved its Springfield Shops just as resourceful. Throughout the late 1930's and into the war the road adroitly played musical chairs with the wheel arrangements on its aging roster. The game began with the most powerful nonarticulated locomotives on the property, a large batch of 60-inch-drivered Baldwin 2-10-2's delivered on the eve of America's entry into World War I. Drag engines, pure and simple. Their boilers proved the inspiration for a totally different breed of power, a heavy 70-inch-drivered 4-8-2 that lost little ground in initial tractive effort (and actually had more if you count the booster), yet could keep hotshots rolling at a pace that with a Santa Fe type would have kinked every last rail in the Ozarks. Frisco produced 11 such 2-10-2-into-4-8-2 rebuilds in 1936, turned out another 23 in 1939. Similarly, in 1937 and 1940-1941 Springfield reconstructed 10 elderly 2500-horsepower Pacifics into streamstyled 3600-horsepower Hudsons fitted with enlarged fireboxes, siphons, boosters, and feedwater heaters. Later, during World War II, the road rebuilt 2-8-0's into Mikes.

Both Frisco and Mopac settled for the 4-8-4 when finances permitted and wartime traffic demanded new locomotives. The Northern proved herself a splendid ridge-runner, an engine that could handle singly an 18-car *Meteor* for the Frisco, say, or sit down and lug on a helper grade such as Rolla Hill. Frisco's 25 Baldwin 4-8-4's in the 4500 series were similar to Soo's Northerns; on test they showed smooth operation at 80 miles per hour with the cutoff as short as 22 per cent. Mopac ordered brand-new power too late in the war to originate fresh blueprints, so it settled for a mild modification of Rio Grande's M-68 class 4-8-4. The 2201-2215 series engines developed a bit less tractive effort than Frisco's 4500's but were possessed of more generous grate-area and heating-surface statistics.

St. Louis Southwestern followed the example of its neighbors by rehabilitating its motive power during bankruptcy, and moreover left the courts with its bills paid in full! Just prior to the depression Cotton Belt had purchased 10 Baldwin 4-8-4's to supplement 2-8-0's, previously the road's heaviest power. These Northerns — delivered in Nile green paint and fitted with chromium cylinder head covers — introduced their wheel arrangement in the Southwest more than a decade before Frisco and Mopac adopted them. Much impressed by their performance, Cotton Belt's trustees ordered 5 more in 1937 from the road's Pine Bluff (Ark.) Shops; and 5 additional homemade sisters followed in 1942-1943. These engines, together with 12 secondhand 4-8-2's purchased from Florida East Coast and Rock Island, were instrumental in Cotton Belt's recovery and its performance during the war. SSW, no less than N&W, made much of the fact that Pine Bluff had built its largest power (to quote the release, "giant locomotives planned, designed, and constructed by the Cotton Belt's own staff of mechanical engineers, mechanical officers, foremen, and workers in company shops"), and the reward was an obvious boost in morale.

Perhaps *the* engine event in modern times in the Southwest, however, was the application of a notably successful wheel arrangement, and it took place on the very solvent Texas & Pacific prior to the financial panic. Technically, Santa Fe originated the 2-10-4 in 1919 when it experimentally inserted a 4-wheel trailing truck under an orthodox Santa Fe type, No. 3829, but the net effect was an exercise in weight distribution techniques — nothing more. In 1925 the real thing took place. T&P's I-class engine was simply an elongated version of builder Lima's famous A-1 Super-Power 2-8-4. Both the 2-8-4 and the 2-10-4 had in common generous 100-square-foot grate areas, long valve travel, articulated 4-wheel trailing trucks, tandem rods, limited cutoff, boosters, feedwater heaters, front-end throttles — all the traits upon which Lima focused. Overnight the beetle-browed Texas engines brought T&P locomotive development in the region from a rear-guard position to the head of the class. For years on end these 70 2-10-4's constituted an oasis of modernism in the Southwest. Unlike many of the original Lima Berks, T&P's 600's never underwent drastic surgery. During the war they were recounterbalanced for speeds up to 60 miles per hour and fitted with steam and signal lines for passenger service. Otherwise the I's hauled T&P through thick and thin with no apologies asked or needed.

R. S. Plummer.

Incidentally, the trailing truck of one T&P 2-10-4, No. 632, was lengthened to accommodate *two* boosters developing 28,200 pounds tractive effort for a grand total of 112,800 pounds with the main cylinders. Which was asking too much of even a Lima boiler.

Lima Super-Power enjoyed scant popularity on Southwest roads, aside from T&P. A commendable exception was Kansas City Southern. In 1937 KCS purchased 10 Texas engines from Lima, the first locomotives with staybolted boilers to carry a pressure as high as 310 pounds per square inch. These 900's explain much about the extraordinary progress made in steam power design during the 1930's. T&P's pioneer 2-10-4's weighed 448,000 pounds and their 63-inch drivers exerted 83,000 pounds tractive effort. Only moderately heavier, the 509,000-pound KCS engines developed 93,300 pounds tractive effort despite 70-inch drivers.

The final years of steam's reign in the Southwest, then, found the region no longer emulating the locomotive practice of others but striking out boldly on its own. One could see as much beside the tower of St. Louis Union Station. There the Eastern engines that had once stirred a nation, Pennsy's K4 and Central's Hudson, were dwarfed by the huge, well-proportioned 4-8-2's and 4-8-4's of the one-time Gould roads. The rising discrepancy in size between the power assigned to, say, the *Southwestern* and the *Sunshine Special* must have prompted interesting conversation between cabs. I

William K. Barham.

REGAL RESURRECTION OF A SANTA FE

A Santa Fe, or 2-10-2, could pull all outdoors — but not very fast. So Frisco's Springfield (Mo.) Shops rebuilt a batch of old 60-inch-drivered World War I era drag engines of that wheel arrangement into 70-inch 4-8-2's in 1936 and 1939. Disc-drivered 4413 explains what all the fuss was about as she approaches the Oklahoma line at Neosho, Mo., at 50 with *Oklahoma Flash* No. 31.

IN TIME FOR TANK TRAINS

In the war the potential bottleneck of the Frisco was the 282-mile, sawtooth-profile Eastern Division which funneled traffic from the Southwest at Monett, Mo., into the St. Louis gateway. To conquer the hills the road bought 25 Baldwin 4-8-4's in 1942-1943 (of specifications remarkably similar to Soo's O-20's). The 4500's could sustain 5600 horsepower at 47 to 65 miles per hour.

ST. LOUIS-SAN FRANCISCO 4400 MOUNTAIN (4-8-2) FOR FREIGHT SERVICE

Series	4400-4422
Cylinders, diameter and stroke	29 x 32 in.
Driving wheel diameter	70 in.
Boiler pressure	210 lbs. per sq. in.
Grate area	76.2 sq. ft.
Evaporative heating surface	5122 sq. ft.
Superheater heating surface	1726 sq. ft.
Weight on drivers	278,950 lbs.
Weight on lead truck	75,180 lbs.
Weight on trailing truck	65,070 lbs.
Weight of total engine	419,200 lbs.
Weight of tender	255,890 lbs.
Tender capacity	5000 gals. oil or 20 tons coal; 14,000 gals. water
Tractive force, engine	68,600 lbs.
Tractive force, booster	8750 lbs.
Builder, date	Baldwin, 1916-1917; rebuilt Springfield, 1939

ST. LOUIS-SAN FRANCISCO 4500 4-8-4 FOR DUAL SERVICE

Series	4503-4514
Cylinders, diameter and stroke	28 x 31 in.
Driving wheel diameter	74 in.
Boiler pressure	255 lbs. per sq. in.
Grate area	88.3 sq. ft.
Evaporative heating surface	4766 sq. ft.
Superheater heating surface	1508 sq. ft.
Weight on drivers	279,630 lbs.
Weight on lead truck	78,900 lbs.
Weight on trailing truck	103,970 lbs.
Weight of total engine	462,500 lbs.
Weight of tender	341,300 lbs.
Tender capacity	24 tons; 18,000 gals.
Tractive force, engine	71,200 lbs.
Builder, date	Baldwin, 1942

ONE OF THE FEW POST-1930 2-8-2'S

Louisiana & Arkansas's 561-565 series Mikes were exactly what they seemed to be: latter-day light U.S.R.A. 2-8-2's modernized in appearance or performance with higher boiler pressure, pumps on pilot beam, and a Delta trailer. They were among the few Mikes built after 1930. L&A train No. 77 departs from Shreveport, La., for New Orleans in 1947 behind the well-kept 565.

Harold K. Vollrath.

OZARK RIDGE-RUNNERS

St. Louis Southwestern managed to climb out of bankruptcy without owing anyone a penny, and helping mightily in the feat were 20 handsome 4-8-4's. Baldwin built the first 10; the balance were homemade in the Pine Bluff (Ark.) Shops. No. 815, rolling west through Texarkana, typifies a class which SSW employees built and loved.

R. S. Plummer.

Collection of George Sennhauser.

LOUISIANA & ARKANSAS E4HT54.8 MIKADO (2-8-2) FOR FREIGHT SERVICE

Series	563-565	Weight on lead truck	38,100 lbs.
Cylinders, diameter and stroke	23 x 32 in.	Weight on trailing truck	52,000 lbs.
Driving wheel diameter	63 in.	Weight of total engine	298,200 lbs.
Boiler pressure	240 lbs. per sq. in.	Weight of tender, loaded	188,600 lbs.
Grate area	60.1 sq. ft.	Tender capacity	4000 gals. oil; 9500 gals. water
Evaporative heating surface	3913 sq. ft.		
Superheater heating surface	1016 sq. ft.	Tractive force, engine	54,800 lbs.
Weight on drivers	208,100 lbs.	Builder, date	Lima, 1936

ST. LOUIS SOUTHWESTERN L1 4-8-4 FOR FREIGHT SERVICE

Series	815-819	Weight on lead truck	80,200 lbs.
Cylinders, diameter and stroke	26 x 30 in.	Weight on trailing truck	97,300 lbs.
Driving wheel diameter	70 in.	Weight of total engine	425,500 lbs.
Boiler pressure	250 lbs. per sq. in.	Weight of tender, loaded	312,000 lbs.
Grate area	88.3 sq. ft.	Tender capacity	5000 gals. oil; 15,000 gals. water
Evaporative heating surface	4724 sq. ft.		
Superheater heating surface	1962 sq. ft.	Tractive force, engine	61,564 lbs.
Weight on drivers	248,000 lbs.	Builder, date	Pine Bluff, 1942-1943

Rail Photo Service: C. T. Wood.

BLUEPRINTS FROM ANOTHER GOULD ROAD

Once upon a time in the Gould fraternity Missouri Pacific and Rio Grande were in the same camp, so it was like old times in 1943 when Mopac ordered 15 4-8-4's to the design of D&RGW's M-68 class. Minus vestibule cab and with a different feedwater heater — which made the 2200's handsomer than the prototype. The 2215 has westbound reefers in the hole at Sandy Hook, Mo.

FROM SUPER-POWER TO SUPER, PERIOD

The Missouri Pacific Northern (No. 2117) getting a wheel on Second 2 out of California, Mo., was originally one of 25 Lima Super-Power Berkshires built in 1930. The new breed of engines steamed fine but their gait was restricted by a 63-inch wheel, a drawback resolved in 1940-1942 when Sedalia Shops mounted the boilers on new cast-steel frames, made 75-inch-drivered Northerns.

Rail Photo Service: C. T. Wood.

MISSOURI PACIFIC 2101 NORTHERN (4-8-4) FOR DUAL SERVICE

Series	2101-2125
Cylinders, diameter and stroke	28 x 30 in.
Driving wheel diameter	75 in.
Boiler pressure	250 lbs. per sq. in.
Grate area	88.3 sq. ft.
Evaporative heating surface	4837 sq. ft.
Superheater heating surface	1953 sq. ft.
Weight on drivers	269,360 lbs.
Weight on lead truck	70,000 lbs.
Weight on trailing truck	96,590 lbs.
Weight of total engine	445,950 lbs.
Weight of tender, 2/3 load	323,880 lbs.
Tender capacity	20 tons; 17,250 gals.
Tractive force, engine	66,640 lbs.
Builder, date	Lima, 1930; rebuilt Sedalia, 1940-1942

MISSOURI PACIFIC 2201 NORTHERN (4-8-4) FOR FREIGHT SERVICE

Series	2201-2215
Cylinders, diameter and stroke	26 x 30 in.
Driving wheel diameter	73 in.
Boiler pressure	285 lbs. per sq. in.
Grate area	106 sq. ft.
Evaporative heating surface	5306 sq. ft.
Superheater heating surface	2200 sq. ft.
Weight on drivers	279,400 lbs.
Weight on lead truck	94,115 lbs.
Weight on trailing truck	115,485 lbs.
Weight of total engine	489,000 lbs.
Weight of tender, 2/3 load	287,700 lbs.
Tender capacity	20 tons; 19,350 gals.
Tractive force, engine	67,200 lbs.
Builder, date	Baldwin, 1943

FOR A BIG STATE, A BIG ENGINE

Never was a wheel arrangement more aptly named than the Texas. Texas & Pacific began building a fleet of 70 such 2-10-4's in 1925 on the heels of Lima's revolutionary A-1 2-8-4. Compared with the 2-10-2's they replaced, the I-class engines hauled 44 per cent 33 per cent faster. Sectionmen step back to admire a T&P Texas in West Texas as the 2-10-4 comes charging up bad Baird Hill.

LIMA AT ITS FINEST

Kansas City Southern's 10 J-class 2-10-4's of 1937 had the first staybolted boilers with 310 pounds pressure. Built to replace doubleheaded 2-8-0's or Mallets on hotshots across the rough 432.4-mile Northern Division between Kansas City and De Queen, Ark., the 70-inch-drivered Texas engines were allowed up to 60 miles per hour on hotshot 77, could take 1800 tons up 1.8 per cent.

Collection of George Sennhauser.

TEXAS & PACIFIC I-1 TEXAS (2-10-4) FOR FREIGHT SERVICE

Series	600-609
Cylinders, diameter and stroke	29 x 32 in.
Driving wheel diameter	63 in.
Boiler pressure	250 lbs. per sq. in.
Grate area	100 sq. ft.
Evaporative heating surface	5113 sq. ft.
Superheater heating surface	2100 sq. ft.
Weight on drivers	300,000 lbs.
Weight on lead truck	41,800 lbs.
Weight on trailing truck	106,200 lbs.
Weight of total engine	448,000 lbs.
Weight of tender, loaded	275,200 lbs.
Tender capacity	5000 gals. oil; 14,000 gals. water
Tractive force, engine	83,000 lbs.
Tractive force, booster	13,000 lbs.
Builder, date	Lima, 1925

KANSAS CITY SOUTHERN J TEXAS (2-10-4) FOR FREIGHT SERVICE

Series	900-904
Cylinders, diameter and stroke	27 x 34 in.
Driving wheel diameter	70 in.
Boiler pressure	310 lbs. per sq. in.
Grate area	107 sq. ft.
Evaporative heating surface	5154 sq. ft.
Superheater heating surface	2075 sq. ft.
Weight on drivers	350,000 lbs.
Weight on lead truck	50,600 lbs.
Weight on trailing truck	108,400 lbs.
Weight of total engine	509,000 lbs.
Weight of tender, loaded	348,000 lbs.
Tender capacity	4500 gals. oil; 21,000 gals. water
Tractive force, engine	93,300 lbs.
Builder, date	Lima, 1937

Collection of George Sennhauser.

Richard Steinheimer.

An Alco 4-8-4, No. 842, forwards the U. S. Mail for Union Pacific through a canyon a few miles east of Green River in a desert setting of pure Wyoming.

THE WEST

. . . for a land of limitless miles and high altitudes there were locomotives to match

THE largest, most audacious steam locomotive designs were developed in the West. Expectedly so, too, for there the grades, curvature, and distances demanded big, often unorthodox power and clearances permitted its construction. Only in the West did a railroad dare to specify a 74-inch driving wheel diameter for a 10-coupled freight locomotive . . . or provide an enormous 182-square-foot grate for consumption of strip-mine coal . . . or reverse a wheel arrangement so that the crew rode ahead of the boiler instead of behind. The engines produced in such quantities for the rest of the U. S. were no match for the mountains. For example, not a single 4-6-4 or 2-8-4 was ever purchased new for service west of the Continental Divide. Operating conditions brought about a conformity of interest for a larger tandem. Every transcontinental but two settled on the simple articulated for freight service, and *all* of them coupled their varnish to the 4-8-4.

The 4-8-4 was very much a Western engine. The original Northerns, delivered in 1926, were bought by Northern Pacific to haul up to nine cars unassisted over a 2.2 per cent ruling grade between Livingston and Missoula, Mont. From that auspicious beginning the wheel arrangement steadily gained ground. As the 4-8-4's grew larger (from an engine weight of 426,000 pounds for NP's 2600 of 1926 to 510,150 pounds for Santa Fe's 2900 of 1944) they broke one record after another.

Distance? Northern Pacific employed its Baldwins on 18-car *North Coast Limiteds* between St. Paul and Missoula, Mont., 1008 miles, without change, the longest coal-burning locomotive run in the world. The oil-fired Northerns of Santa Fe topped that feat by first handling the *Chief* from La Junta, Colo., to Los Angeles, 1235 miles, without relief and subsequently taking on the 1765-mile Kansas City-L.A. haul via Amarillo.

Speed? In 1938 the Association of American Railroads set out to probe the limits of steam power with a 16-car, 1000-ton test train. A 14-year-old Pennsy K4 managed 91 miles per hour with this train; and a 7-foot-drivered North Western 4-6-4 got it up to 95. But on October 11 a year-old, 77-inch-drivered Union Pacific Northern, No. 815, ran the special up to 102 miles per hour on a slightly descending 0.13 per cent grade between Grand Island and Omaha, Nebr.

Capacity? Santa Fe worked its 80-inch-drivered 4-8-4's up grades as stiff as Raton's 3.5 per cent; Rio Grande once scheduled its M-68's on a 177-mile nonstop haul. Or consider Southern Pacific's now legendary Lima GS engines. They began life on the glamorous *Daylights*, keeping a relatively heavy streamliner to time in mountainous territory by day and handling expedited merchandise hotshots at night; in the war they roamed the system from El Paso to Portland, performing herculean feats of mileage with troop trains and overloaded conventional limiteds; and they wound up their days in an exacting, 38-mile commutation service between San Francisco and San Jose, Calif.

Santa Fe had this in common with Pennsylvania. Following an early and exhaustive study of several species of the articulated, both roads swore off jointed power for good. Santa Fe rebuilt its last Mallets into 2-10-2's in World War I and didn't operate any more until World War II, when pressure of war traffic forced purchase of eight old N&W Y3's for helper duty. The road stuck with huge 2-10-2's for tonnage on the same grades (and the same track in Cajon and Tehachapi passes) for which rivals Espee and UP specified articulateds. After having technically originated the 2-10-4 in 1919 (a 2-10-2 fitted with a 4-wheel trailing truck), Santa Fe bought a "legitimate" Texas in 1930 — No. 5000. Finally, in 1938, the road began its famous 5001 class. The big Baldwins were a locomotive landmark. Imagine a 10-coupled, 269-ton freight engine rated at 93,000 pounds tractive effort — *with 74-inch drivers!* Larger drivers, indeed, than many a road ever managed on strictly passenger power. On a wartime test over the Missouri Division between Argentine, Kans., and Shopton, Ia., No. 5011 demonstrated once and for all the possibilities of nonarticulated power. Rolling east with a 94-car, 5441-ton train she hit the bottom of Cardy Hill at 56 miles per hour, went over the summit at 18 after 5 miles of mostly 0.8 per cent ascending grade.

Other Western roads agreed to disagree with Atchison on the articulated principle. The most demonstrative, Union Pacific, returned to jointed engines, in fact, after displacing slow-footed Mallets with fleets of triple-cylindered 4-10-2's and 4-12-2's. In 1936 UP set out to equal the boiler output of its renowned 12-coupled 9000's in a wheel arrangement that would permit higher speeds as well as freedom from the restrictions of a long rigid frame and an inside cylinder. The Challenger 4-6-6-4 admirably filled the bill. The design was a runaway favorite, far and away the most popular of latter-day articulateds, and it quickly moved onto the rosters of not only Western transcontinentals (notably Northern Pacific) but roads as far East as Delaware & Hudson and Western Maryland. A 4-6-6-4 could and did work heavy passenger trains with no apology, and in the words of one expert it rode as well at 60 to 70 miles per hour as any articulated ever built.

For grades and tonnages with which only 16 drivers could cope, there were as many answers forthcoming as there were railroads to buy such huge machines. Great Northern and Western Pacific stuck to the 2-8-8-2 for mountain service; GN's homemade R-2 excelled all other simple articulateds with a tractive effort of 146,000 pounds and WP's Baldwins became, with only minor blueprint modification, Missabe's capable Yellowstones. Southern Pacific favored the 2-8-8-4, running the wheel arrangement as such in the case of 12 coal-burning AC-9's by Lima and "flopping" the type into a cab-forward oil-burner in the creation of 195 4-8-8-2's delivered by Baldwin between 1928 and 1944. Espee's unorthodox mountain climbers constituted the only instance of employment of a 4-wheel engine or guide truck to support a large firebox. It is enough to say of the 4100's and 4200's that without them the railroad would have broken under its immense war load with disastrous effects.

The "world's largest" accolade has always carried a peculiar appeal for locomotive men — or it did before the limitless multiple-unit connection of diesels made the title academic. Northern Pacific grabbed the glory in 1928 when Alco completed a 357½-ton 2-8-8-4 demonstrator. The task for No. 5000 to tackle was the 216-mile Mandan (N. Dak.)-Glendive (Mont.) run. Because of the 1.1 per cent grades NP had been obliged to break up 4000-ton trains arriving at each terminal and dispatch them across the intervening badlands in 2225-ton hauls — the largest tonnage the Mikes could tote. The 5000 went it alone, cut train-miles almost in half. And because she burned "Rosebud" lignite, the 2-8-8-4 had the largest grate applied before or since: 182 square feet! NP was satisfied, and in 1930 Baldwin delivered 11 slightly heavier sisters. Upgraded in the war with new frames and roller bearings, these Z-5's enjoyed a long and useful life.

Wendell Mortimer.

Union Pacific made the most and last use of the "largest" label in 1941 when it sent 20 simple 4-8-8-4 Big Boys (a named chalked on the smokebox of No. 4000 at the Schenectady [N.Y.] plant of Alco) to work on the Ogden-Green River (Wyo.) run. Whether he was observed climbing Wasatch or Sherman, Big Boy was all his nickname connoted and a fine example of the immense strides taken in articulated locomotives since their design was temporarily frustrated in the abortive, awkward Triplex compounds of one world war earlier. Wide open, a 4000 consumed up to 22 tons of coal and 100,000 pounds of water an hour and produced in one authenticated dynamometer-car test 6290 drawbar horsepower at 41.4 miles per hour. That such a vast steam-generating plant could be mounted, not on concrete and clothed in brick, but on a flexible base which rode comfortably at 50 miles per hour, and that the output of its boiler could be so harnessed that the machine could lift up to 4200 tons up a 1.14 per cent gradient at 18 to 20 miles per hour — these circumstances will be remarked upon as long as men gather to talk of steam and steel. I

Otto C. Perry.

DUAL-PURPOSE 4-6-6-4'S

Undeterred by 3 per cent grades and 16-degree curves, Rio Grande assaulted its tough profile in 1938 with 10 L-105 class 4-6-6-4's, added 5 more engines to the series in 1941. On continuous 2 per cent an L-105 with a helper could keep 2850 tons moving 18 miles per hour. The 3703 demonstrates her varnish ability with a westbound troop train up Tennessee Pass.

R. H. Kindig.

FOR A RECOVERING RIO GRANDE

Denver & Rio Grande Western, commonly regarded as the toughest railroad to operate in the U. S., sparked its depression overhaul with big Baldwins. No. 1801, getting out of Denver with the *Scenic Limited* in 1938, was one of five M-68 4-8-4's in the program. They once worked nonstop over the 177 miles between Helper, Utah, ahd Grand Junction, Colo., with train No. 6.

DENVER & RIO GRANDE WESTERN M-68 NORTHERN (4-8-4) FOR DUAL SERVICE

Series	1800-1804	Weight on lead truck	85,837 lbs.
Cylinders, diameter and stroke	26 x 30 in.	Weight on trailing truck	114,351 lbs.
Driving wheel diameter	73 in.	Weight of total engine	479,360 lbs.
Boiler pressure	285 lbs. per sq. in.	Weight of tender, loaded	394,000 lbs.
Grate area	106 sq. ft.	Tender capacity	26 tons; 20,000 gals.
Evaporative heating surface	5506 sq. ft.	Tractive force, engine	67,200 lbs.
Superheater heating surface	2336 sq. ft.	Builder, date	Baldwin, 1938
Weight on drivers	279,172 lbs.		

DENVER & RIO GRANDE WESTERN L-105 CHALLENGER (4-6-6-4) FOR FREIGHT SERVICE

Series	3700-3709	Weight on lead truck	84,550 lbs.
Cylinders, diameter and stroke	23 x 32 in. (4)	Weight on trailing truck	119,411 lbs.
Driving wheel diameter	70 in.	Weight of total engine	641,900 lbs.
Boiler pressure	255 lbs. per sq. in.	Weight of tender, loaded	394,000 lbs.
Grate area	136.5 sq. ft.	Tender capacity	26 tons; 20,000 gals.
Evaporative heating surface	6341 sq. ft.	Tractive force, engine	105,000 lbs.
Superheater heating surface	2628 sq. ft.	Builder, date	Baldwin, 1937
Weight on drivers	479,939 lbs.		

Collection of H. L. Broadbelt.

Fred H. Matthews Jr.

Henry R. Griffiths Jr.

WHEN WESTERN PACIFIC CAME OF AGE

Western Pacific conquered the Sierra Nevada in 1909 with its famous Feather River Canyon line that never exceeds 1 per cent of gradient, but it was not until 1931 that WP came of age. Big-time railroading was born at Baldwin in the shape of 6 massive 2-8-8-2's (supplemented by 4 more in 1938). Extra 259 East is approaching Williams Loop in the canyon at a point east of Massack.

"MOST POWERFUL" SIMPLE ARTICULATEDS

Big Boy was heavier, all right, but no other simple articulated equaled the 146,000-pound tractive effort of Great Northern's R-2 class 2-8-8-2's — homemade in 1929-1930. They were also the biggest power with Belpaire boilers. In September 1940, No. 2048 brings a 65-car freight west out of a snowshed on the east slope of Marias Pass against a backdrop of Glacier Park mountains.

GREAT NORTHERN R-2 2-8-8-2 FOR FREIGHT SERVICE

Series	2044-2059
Cylinders, diameter and stroke	28 x 32 in. (4)
Driving wheel diameter	63 in.
Boiler pressure	240 lbs. per sq. in.
Grate area	126 sq. ft.
Evaporative heating surface	7947 sq. ft.
Superheater heating surface	3515 sq. ft.
Weight on drivers	544,000 lbs.
Weight on lead truck	37,750 lbs.
Weight on trailing truck	49,000 lbs.
Weight of total engine	630,750 lbs.
Weight of tender, loaded	372,780 lbs.
Tender capacity	5800 gals. oil; 22,000 gals. water
Tractive force, engine	146,000 lbs.
Builder, date	Hillyard, 1929-1930

WESTERN PACIFIC 257 2-8-8-2 FOR FREIGHT SERVICE

Series	257-260
Cylinders, diameter and stroke	26 x 32 in. (4)
Driving wheel diameter	63 in.
Boiler pressure	235 lbs. per sq. in.
Grate area	145 sq. ft.
Evaporative heating surface	6811 sq. ft.
Superheater heating surface	2152 sq. ft.
Weight on drivers	549,656 lbs.
Weight on lead truck	48,134 lbs.
Weight on trailing truck	65,310 lbs.
Weight of total engine	663,100 lbs.
Weight of tender, loaded	403,350 lbs.
Tender capacity	6000 gals. oil; 22,000 gals. water
Tractive force, engine	137,000 lbs.
Tractive force, booster	13,900 lbs.
Builder, date	Baldwin, 1938

Collection of H. L. Broadbelt.

Henry R. Griffiths Jr.

Henry R. Griffiths Jr.

BIGGEST OF ALL . . . BIG BOY

Union Pacific's world-renowned 4000-series Big Boys were the heaviest locomotives produced in steam's finest hour and the only engines of their 4-8-8-4 wheel arrangement. In addition to moving tonnage between Cheyenne and Ogden the 25 Alcos earned UP enormous press, radio, and TV coverage. Extra 4007 East has 100 cars rolling 35 miles an hour between Hermosa and Dale Creek.

UP INVENTED THE TYPE AND THE NAME

Far and away the most popular articulated after its development in 1936 was the Challenger, or 4-6-6-4. Union Pacific invented and named the high-speed articulated and owned more (105 engines) than any other road. The oil-burning 3701 is shoving hard on the caboose end of eastbound tonnage in Echo Canyon on the climb to Wahsatch after a pause for water in Echo City, Utah.

Collection of Arthur Stensvad.

UNION PACIFIC 4664-3 CHALLENGER (4-6-6-4) FOR FREIGHT SERVICE

Series	3950-3969	Weight on lead truck	101,000 lbs.
Cylinders, diameter and stroke	21 x 32 in. (4)	Weight on trailing truck	122,000 lbs.
Driving wheel diameter	69 in.	Weight of total engine	627,000 lbs.
Boiler pressure	280 lbs. per sq. in.	Weight of tender, 2/3 load	348,500 lbs.
Grate area	132.2 sq. ft.	Tender capacity	28 tons; 25,000 gals.
Evaporative heating surface	4795 sq. ft.	Tractive force, engine	97,350 lbs.
Superheater heating surface	2162 sq. ft.	Builder, date	American, 1942
Weight on drivers	404,000 lbs.		

UNION PACIFIC 4884-1 "BIG BOY" (4-8-8-4) FOR FREIGHT SERVICE

Series	4020-4024	Weight on lead truck	99,800 lbs.
Cylinders, diameter and stroke	23¾ x 32 in. (4)	Weight on trailing truck	127,200 lbs.
Driving wheel diameter	68 in.	Weight of total engine	772,000 lbs.
Boiler pressure	300 lbs. per sq. in.	Weight of tender, 2/3 load	348,000 lbs.
Grate area	150.3 sq. ft.	Tender capacity	28 tons; 25,000 gals.
Evaporative heating surface	5755 sq. ft.	Tractive force, engine	135,375 lbs.
Superheater heating surface	2043 sq. ft.	Builder, date	American, 1944
Weight on drivers	545,000 lbs.		

R. H. Kindig.

Robert Hale.

ONE OF THE ALL-TIME GREATS

Who but Santa Fe would have had the audacity to specify a 74-inch wheel for a 10-coupled locomotive? Yet so large were the boilers that the 35 5001-class engines exerted a tractive effort almost equal to certain modern articulateds, and at 40 miles per hour they could produce 5600 drawbar horsepower. Extra 5017 West near Tejon, N. Mex., is working on 83 cars.

DEAR TO THE HEART OF SANTA FE

Few steam locomotives were so justifiably acclaimed as Santa Fe's 65 Northerns. Huge and tireless creatures, they could and did cover the 1765 miles between Kansas City and L.A. without change, conquer Raton's 3 per cent, and exceed 100 miles per hour. The 2915, a wartime Baldwin, has her stack extension up with the third *Grand Canyon* westbound at Frost, Calif.

ATCHISON, TOPEKA & SANTA FE 2900 NORTHERN (4-8-4) FOR DUAL SERVICE

Series	2900-2929	Weight on lead truck	93,820 lbs.
Cylinders, diameter and stroke	28 x 32 in.	Weight on trailing truck	122,470 lbs.
Driving wheel diameter	80 in.	Weight of total engine	510,150 lbs.
Boiler pressure	300 lbs. per sq. in.	Weight of tender, 2/3 load	369,690 lbs.
Grate area	108 sq. ft.	Tender capacity	7000 gals. fuel; 24,500 gals. water
Evaporative heating surface	5312.5 sq. ft.	Tractive force, engine	66,000 lbs.
Superheater heating surface	2366 sq. ft.	Builder, date	Baldwin, 1943-1944
Weight on drivers	293,860 lbs.		

ATCHISON, TOPEKA & SANTA FE 5001 TEXAS (2-10-4) FOR FREIGHT SERVICE

Series	5001-5010	Weight on lead truck	49,920 lbs.
Cylinders, diameter and stroke	30 x 34 in.	Weight on trailing truck	123,660 lbs.
Driving wheel diameter	74 in.	Weight of total engine	545,260 lbs.
Boiler pressure	310 lbs. per sq. in.	Weight of tender, loaded	359,900 lbs.
Grate area	121.5 sq. ft.	Tender capacity	23 tons; 21,000 gals.
Evaporative heating surface	6075 sq. ft.	Tractive force, engine	93,000 lbs.
Superheater heating surface	2675 sq. ft.	Builder, date	Baldwin, 1937
Weight on drivers	371,680 lbs.		

Collection of H. L. Broadbelt.

R. V. Nixon.

HEAVIEST CHALLENGERS EVER BUILT

American Locomotive Company, which introduced the 4-6-6-4 (on Union Pacific) in 1936, constructed more Challengers than any other builder and, in Northern Pacific's Z-8 class, built the heaviest. NP owned 48 4-6-6-4's in three classes. The latest, most modern series is exemplified by Extra 5141 West at Missoula, Mont., in April 1943. Uncommonly large grate areas were for lignite.

BIRTHPLACE OF THE 4-8-4

Northern Pacific pioneered the Northern in late 1926, later purchased the famous Timken Four Aces 4-8-4, went on to build a modern fleet of Baldwins — the newest of which ran the 18-car *North Coast Limited* St. Paul-Livingston, Mont. (the longest coal-burning locomotive run without change). The 2650, an A-2, is on a main train east of Bozeman, Mont., within Rocky Canyon walls.

NORTHERN PACIFIC A-2 NORTHERN (4-8-4) FOR PASSENGER SERVICE

Series	2650-2659	Weight on lead truck	96,600 lbs.
Cylinders, diameter and stroke	28 x 31 in.	Weight on trailing truck	113,000 lbs.
Driving wheel diameter	77 in.	Weight of total engine	489,400 lbs.
Boiler pressure	260 lbs. per sq. in.	Weight of tender, loaded	387,600 lbs.
Grate area	115 sq. ft.	Tender capacity	27 tons; 20,000 gals.
Evaporative heating surface	4964.3 sq. ft.	Tractive force, engine	69,800 lbs.
Superheater heating surface	2174 sq. ft.	Builder, date	Baldwin, 1934
Weight on drivers	279,800 lbs.		

NORTHERN PACIFIC Z-8 CHALLENGER (4-6-6-4) FOR FREIGHT SERVICE

Series	5130-5149	Weight on lead truck	78,000 lbs.
Driving wheel diameter	70 in.	Weight on trailing truck	122,000 lbs.
Boiler pressure	260 lbs. per sq. in.	Weight of total engine	644,000 lbs.
Grate area	152.3 sq. ft.	Weight of tender, 2/3 load	349,800 lbs.
Evaporative heating surface	5749 sq. ft.	Tender capacity	27 tons; 25,000 gals.
Superheater heating surface	2105 sq. ft.	Tractive force, engine	106,900 lbs.
Weight on drivers	444,000 lbs.	Builder, date	American, 1943-1944

J. F. Orem.

MOST DISTINCTIVE ARTICULATEDS BUILT

Extra 4223 East near Truckee, Calif., in the Sierra Nevada typifies the 4-8-8-2 wheel arrangement that Southern Pacific originated in 1928 by flopping a 2-8-8-4 so that the crew rode on the vision end. Espee's AC engines roamed the system from the Cascade Mountains to the New Mexico desert, handled passenger trains as well as tonnage. All told, Baldwin delivered 195 of the monsters.

FOR THE "MOST BEAUTIFUL" TRAIN

Californians brook no argument on the subject: for them the Southern Pacific's *Coast Daylight* was the world's most beautiful streamliner — a band of red, orange, and black drawn by appropriately styled Lima GS 4-8-4's. Sample 4453 has the westbound *Morning Daylight* bound for San Francisco. Lima delivered 60 GS-class engines; the final 10 were stripped-down wartime locomotives.

Martin Litton.

SOUTHERN PACIFIC GS-4 GENERAL SERVICE (4-8-4) FOR PASSENGER SERVICE

Series	4430-4449
Cylinders, diameter and stroke	25½ x 32 in.
Driving wheel diameter	80 in.
Boiler pressure	300 lbs. per sq. in.
Grate area	90.4 sq. ft.
Evaporative heating surface	4887 sq. ft.
Superheater heating surface	2086 sq. ft.
Weight on drivers	275,700 lbs.
Weight on lead truck	81,300 lbs.
Weight on trailing truck	118,000 lbs.
Weight of total engine	475,000 lbs.
Weight of tender, 2/3 load	313,730 lbs.
Tender capacity	5880 gals. oil; 23,300 gals. water
Tractive force, engine	64,760 lbs.
Tractive force, booster	13,000 lbs.
Builder, date	Lima, 1941

SOUTHERN PACIFIC AC-11 ARTICULATED-CONSOLIDATION (4-8-8-2) FOR DUAL SERVICE

Series	4245-4274
Cylinders, diameter and stroke	24 x 32 in. (4)
Driving wheel diameter	63½ in.
Boiler pressure	250 lbs. per sq. in.
Grate area	139 sq. ft.
Evaporative heating surface	6470 sq. ft.
Superheater heating surface	2616 sq. ft.
Weight on drivers	531,700 lbs.
Weight on lead truck	76,400 lbs.
Weight on trailing truck	49,800 lbs.
Weight of total engine	657,900 lbs.
Weight of tender, loaded	393,300 lbs.
Tender capacity	6100 gals. oil; 22,000 gals. water
Tractive force, engine	124,300 lbs.
Builder, date	Baldwin, 1942-1943

JUST MENTION "800" — THAT'S ALL

Along the UP Trail, America's original transcontinental, mention of the 800's will long recall the capable, long-legged Northerns that were built for standard-weight limiteds but often doubled for diesels on the *City* streamliners. Elephant-eared No. 844, last of the lot, displays her best form as she rolls a railfan extra west out of a tunnel in Utah's Echo Canyon in 1948.

R. H. Kindig.

BUILT FROM BORROWED BLUEPRINTS

Western Pacific wisely devoted its design energies to the articulateds which produced most of its gross, left the 4-6-0's that came with the road plus a few secondhand Florida East Coast Mountains for its sparse passenger service. When power got short in World War II WP bought six 4-8-4's built to the blueprints of Espee GS-6. No. 483 heads the *Royal Gorge* into Wendover.

Rail Photo Service: Ben F. Cutler.

WESTERN PACIFIC 481 GENERAL SERVICE (4-8-4) FOR DUAL SERVICE

Series	481-486
Cylinders, diameter and stroke	27 x 30 in.
Driving wheel diameter	73½ in.
Boiler pressure	260 lbs. per sq. in.
Grate area	90.2 sq. ft.
Evaporative heating surface	4922 sq. ft.
Superheater heating surface	2086 sq. ft.
Weight on drivers	280,950 lbs.
Weight on lead truck	73,650 lbs.
Weight on trailing truck	111,500 lbs.
Weight of total engine	466,100 lbs.
Weight of tender, 2/3 load	316,250 lbs.
Tender capacity	6000 gals. fuel; 23,300 gals. water
Tractive force, engine	64,200 lbs.
Tractive force, booster	11,300 lbs.
Builder, date	Lima, 1943

UNION PACIFIC FEF-3 NORTHERN (4-8-4) FOR DUAL SERVICE

Series	835-844
Cylinders, diameter and stroke	25 x 32 in.
Driving wheel diameter	80 in.
Boiler pressure	300 lbs. per sq. in.
Grate area	100.2 sq. ft.
Evaporative heating surface	4294 sq. ft.
Superheater heating surface	1400 sq. ft.
Weight on drivers	270,300 lbs.
Weight on lead truck	100,600 lbs.
Weight on trailing truck	119,800 lbs.
Weight of total engine	490,700 lbs.
Weight of tender	411,400 lbs.
Tender capacity	25 tons; 23,500 gals.
Tractive force, engine	63,800 lbs.
Builder, date	American, 1944

Rail Photo Service: Frank MacKinley.

FOR MANY YEARS THEY WERE THE BIGGEST

In 1928, when Alco built prototype No. 5000, and for years thereafter, Northern Pacific's Yellowstone 2-8-8-4's were the world's largest locomotives. And no other engine ever carried a larger grate (182 square feet!). The 12 engines of the Z-5 class were rebuilt in 1941 with new frames and roller bearings. No. 5003 has the role of helper on the way upgrade from Bozeman, Mont.

COAL BURNERS COULDN'T BE CAB-FORWARDS

Flying green, Lima-built AC-9 3804 brings Second 990 East, a California fruit block, into the Rock Island interchange at Tucumcari, N. Mex., in June 1951. Southern Pacific bought 12 of the articulateds for dual service in 1939, disavowed its usual allegiance to cab-forward power so that the 2-8-8-4's could burn local coal from company mines. The AC-9's wound up on oil in California!

Rail Photo Service: W. G. Fancher.

SOUTHERN PACIFIC AC-9 ARTICULATED-CONSOLIDATION (2-8-8-4) FOR DUAL SERVICE

Series	3800-3811	Weight on lead truck	48,300 lbs.
Cylinders, diameter and stroke	24 x 32 in. (4)	Weight on trailing truck	110,400 lbs.
Driving wheel diameter	63½ in.	Weight of total engine	689,900 lbs.
Boiler pressure	250 lbs. per sq. in.	Weight of tender, 2/3 load	320,800 lbs.
Grate area	139.3 sq. ft.	Tender capacity	28 tons; 22,120 gals.
Evaporative heating surface	6918 sq. ft.	Tractive force, engine	124,300 lbs.
Superheater heating surface	2831 sq. ft.	Builder, date	Lima, 1939
Weight on drivers	531,200 lbs.		

NORTHERN PACIFIC Z-5 YELLOWSTONE (2-8-8-4) FOR FREIGHT SERVICE

Number	5000	Weight on lead truck	48,500 lbs.
Cylinders, diameter and stroke	26 x 32 in. (4)	Weight on trailing truck	115,500 lbs.
Driving wheel diameter	63 in.	Weight of total engine	717,000 lbs.
Boiler pressure	250 lbs. per sq. in.	Weight of tender, loaded	401,000 lbs.
Grate area	182 sq. ft.	Tender capacity	27 tons; 21,200 gals.
Evaporative heating surface	7673 sq. ft.	Tractive force, engine	140,000 lbs.
Superheater heating surface	3219 sq. ft.	Tractive force, booster	13,400 lbs.
Weight on drivers	553,000 lbs.	Builder, date	American, 1928

R. V. Nixon.

WHAT WAS GOOD ENOUGH FOR NP

Since its birth the 945-mile Spokane, Portland & Seattle has avoided expensive custom locomotive design in favor of secondhand power bought from parents Great Northern and Northern Pacific — or new engines built to NP's blueprints. Extra 911 West near Marshall, Wash., is headed by a 4-6-6-4 which, except for the emblem, could pass for an NP Z-8. Which, in effect, it is.

SISTER 3461 MADE A RECORD

In the late 1930's three railroads — Milwaukee, North Western, and Santa Fe — purchased 7-foot-drivered 4-6-4's with 300-pounds-pressure boilers. The six 3460-class Baldwins of Santa Fe normally handled the *Chief* on the Chicago-La Junta (Colo.) run, but in December 1937 the 3461 ran without change from L.A. to Chicago on mail train No. 8. Sister 3462 enters Lawrence, Kans.

Robert B. Olmsted.

ATCHISON, TOPEKA & SANTA FE
3460 HUDSON (4-6-4) FOR PASSENGER SERVICE

Series	3461-3465
Cylinders, diameter and stroke	23½ x 29½ in.
Driving wheel diameter	84 in.
Boiler pressure	300 lbs. per sq. in.
Grate area	98.5 sq. ft.
Evaporative heating surface	4770 sq. ft.
Superheater heating surface	2080 sq. ft.
Weight on drivers	213,440 lbs.
Weight on lead truck	83,950 lbs.
Weight on trailing truck	114,990 lbs.
Weight of total engine	412,380 lbs.
Weight of tender, loaded	396,340 lbs.
Tender capacity	7000 gals. oil; 21,000 gals. water
Tractive force, engine	49,300 lbs.
Builder, date	Baldwin, 1937

SPOKANE, PORTLAND & SEATTLE
Z-8 CHALLENGER (4-6-6-4) FOR FREIGHT SERVICE

Number	911
Cylinders, diameter and stroke	23 x 32 in. (4)
Driving wheel diameter	70 in.
Boiler pressure	260 lbs. per sq. in.
Grate area	152.3 sq. ft.
Evaporative heating surface	5749 sq. ft.
Superheater heating surface	2105 sq. ft.
Weight on drivers	445,500 lbs.
Weight on lead truck	77,500 lbs.
Weight on trailing truck	115,500 lbs.
Weight of total engine	638,500 lbs.
Weight of tender, 2/3 load	351,200 lbs.
Tender capacity	6530 gals. oil; 25,000 gals. water
Tractive force, engine	106,900 lbs.
Builder, date	American, 1944

BEYOND OUR BORDERS

. . . Mexico emulated U.S. practice but Canada frequently pioneered

STRETCHED across the Dominion of Canada are two parallel but dissimilar railways — the continent's only true transcontinentals north of the Rio Grande. The larger road, Canadian National, is publicly owned and operated; it began life in 1922 as what John W. Barriger has termed "a heterogeneous collection of 22,000 miles of bankrupt railroad." Its rival, Canadian Pacific, originally unified Canada with steel rails in 1885, and since then has become the only major privately owned and operated railway outside the U.S. It was to be expected in steam that the differences between CNR and CPR would be reflected in their locomotives.

The role of unifying the insolvents into a truly national system fell to U. S.-born and Pennsy-trained Sir Henry Thornton (who went to Canada via the Long Island and England's Great Eastern). The hundreds of new engines ordered by his administration during the 1920's ranged from Mikes to Mountains to road diesels (Thornton envisioned "wireless electrification" and experimented with a pair of "oil electrics" as early as 1928 in an attempt to equal the performance of a steam 4-8-2). One wheel arrangement stood out, however, then and until dieselization. CNR's Confederation 4-8-4 of 1927 (she missed by seven months being the first of her type to NP) not only was an early expression of the modern engine but was ideally suited to the problems of the recently amalgamated system. The 4-8-4 layout permitted a high-horsepower machine adaptable to either passenger service or manifest freight without exceeding limited axle loadings, and sufficiently modern to cross two or three engine districts without change. Between 1927 and 1944 Canadian National and its U. S. subsidiary Grand Trunk Western acquired more than 200 Northerns — more 4-8-4's, in fact, than Espee, Rio Grande, Santa Fe, and UP could count among them. Of this fleet, 11 were streamlined, the most successfully so being the U-4-a Montreal-built CNR series 6400-6404. The balance of the 4-8-4's varied little in fundamental design but considerably in external appearance. An unusual esthetic trait of a Canadian National 4-8-4 was its heavy, quite formidable appearance while, in point of total weight, the U-class locomotives were lightweights by U. S. standards — approximately equal to, say, Nashville, Chattanooga & St. Louis's Dixies.

Canadian Pacific took another tack. The railway sampled a pair of 4-8-4's in 1928. Homemade, Angus-built Nos. 3100-3101 — "the heaviest and most powerful passenger locomotives in the British Empire" — were ordered to supplement Pacifics. However, the railway's reaction was negative, just as it had been in 1914 to a pair of 4-8-2's. A year later CPR began building a fleet of 75-inch-drivered Hudsons that would eventually number 65 engines. These 4-6-4's were all the more remarkable because they constituted the only major application of the wheel arrangement to dual-

Rail Photo Service: George C. Corey.

The Mike on Canadian Pacific time freight for the coast is given a helping hand west from Revelstoke to Clanwilliam, B. C., 8 miles, by Selkirk 5930.

service work. The first 20 H1's delivered in 1929-1930 were of conventional, taut CPR stamp (*e.g.*, with Elesco heater, centered headlight, vestibule cab), but the next batch in 1937 introduced the railway's famous semistreamlining pattern, keynoted by a smooth jacketing, recessed headlight, solid pilot, and domeless boiler. In 1939 Nos. 2850 and 2851 hauled the Royal and pilot trains, respectively, of King George VI and Queen Elizabeth from Quebec City to Vancouver, 3224 miles; and during a run which involved 25 different engine crews and helper service only in the mountains, the 4-6-4's developed no mechanical ailments. By permission of the British Government the semistreamlined H1's were subsequently called Royal Hudsons and fitted with an embossed crown on the running board scarf plate.

These 4-6-4's worked such long hauls as the 841-mile Winnipeg-Calgary run without change . . . faced the fiercest winter weather on the Continent . . . and piled up notable utilization records (No. 2859, for example, averaged 596 miles per day for eight straight months).

Other surprises on CPR's roster:

¶The railway purchased its first Pacific in 1906, remained infatuated with the 4-6-2 wheel arrangement until dieselization. No less than 195 new Pacifics were added to the roster between 1938 and 1948; the last 73 of these engines were not heavier but far lighter than their predecessors (230,000 pounds vs. 324,000) and were intended for secondary mains and branch lines on which axle loadings were light and existing motive power was elderly. Canadian Pacific operated the 4-6-2 as a dual-purpose locomotive, just as it did the 4-6-4.

¶Whereas the U. S. had merely dabbled with the 4-4-4 (Reading originated the type with an engine it rebuilt into a 4-4-2; and B&O rebuilt a 4-4-2 into a 4-4-4), CPR called it the Jubilee, ordered 5 in 1936 for fast semistreamliners, and bought 20 somewhat lighter and Spartanized 4-4-4's in 1937-1938 for local and branch-line services. Streamstyled in the manner of the Hudsons, these lightweights were unusually capable for their size. Indeed, they constituted one of the rare examples in North America of a road buying new power for secondary services.

¶For the mountain run between Calgary, Alta., and Revelstoke, B. C., through the Spiral Tunnels and Kicking Horse Pass, Canadian Pacific depended upon a fleet of 36 Selkirk 2-10-4's, ordered in lots of 20, 10, and 6 approximately a decade apart beginning in 1929. (One more 2-10-4, No. 8000, was a multi-pressure, 3-cylinder experimental homemade at Angus Shops in 1931, retired in 1940 — and the heaviest steam engine CPR ever owned.) These 5900's were employed on gradients as heavy as 2.2 per cent, worked in helper, freight, and passenger slots, were allowed up to 65 miles per hour despite a 63-inch driving wheel. Though they were exceeded in weight by at least two classes of U. S. 2-8-4's, the T1's were nevertheless stout mountain battlers — and (in the case of Nos. 5920-5935) were the only streamstyled 10-coupled power ever built.

South of the Rio Grande there has always been much to intrigue the student of the steam locomotive, ranging from the Mexicano's celebrated Fairlies clear through to postwar 4-8-4's in modern times. Since 1930 the National Railways of Mexico has purchased high-mounted passenger 4-8-0's, simple 2-6-6-2's, 3-cylinder Pacifics, Hudsons, and Northerns from U. S. and Canadian builders (and these engines have been supplemented by such secondhanders as ex-FEC 4-8-2's, P&LE Mikes, and T&P 2-10-2's).

Perhaps Mexico's chief claim to modern steam locomotive fame was a series of six-coupled articulateds for the 3-foot-gauge line between Buenavista Station, Mexico City, and Toluca, approximately 35 miles. This route was constructed in 1883 and involves a horseshoe curve, two tunnels, ruling grades of 3.76 per cent southbound and 3.91 per cent northbound (with patches up to 4.8 per cent), and severe curvature. To eliminate a helper station on the Toluca line the NdeM bought 6 simple 2-6-6-2's in 1928, 2 more in 1934, and a final 2 in 1937. These 10 articulateds exerted 41,100 pounds tractive effort (compared with 37,100 pounds for the heaviest of Rio Grande's narrow-gauge Mikados), and they were rated at 9900 metric tons on the level and 245 on 3.9 per cent. What with smokebox-mounted pumps, outboard journals on engine and trailing trucks, and occasionally Elesco feedwater heaters, the HR-01's looked (and behaved) like miniaturized Chessie coal haulers. Following the standard-gauging of their Toluca line, all of them were out of service by 1954 and retired soon thereafter.

Everett L. DeGolyer Jr.

NdeM's last new steam power was a series of 32 Alco and Baldwin 4-8-4's delivered in 1946. Among the lightest engines of their wheel arrangement ever built, the Niagras were nevertheless modern in detail, balance, and appearance. Their effective service life was cut short by an almost simultaneous switch to diesels — a circumstance shared with many an engine north of the border.

Thus it was that Canada was extraordinarily individualistic in steam locomotive design whereas Mexico was happy to accept a scaled-down version of U. S. practice. In an esthetic sense, though, who was to choose between a Selkirk bringing the *Dominion* west into Kicking Horse Pass or an HR-01 flailing her Walschaerts over the summit at La Cima?

Robert Hale.

CPR BURIED THE HEADLIGHT

A trait of latter-day steam power on Canadian Pacific was the recessed headlight, illustrated on the engine of Extra 2380 East running between Swift Current and Moose Jaw, Sask., in December 1956. Possessed of many thinly trafficked lines and dual-purpose assignments for engines, CPR commonsensically employed Pacifics as all-purpose engines, bought 4-6-2's until dieselization.

THE BEST WAS NOT ALWAYS THE BIGGEST

Reading originated the 4-4-4 wheel arrangement (on an experimental that later became a 4-4-2) but Canadian Pacific dubbed it "Jubilee" in 1936 (in honor of 50 years of coast-to-coast passenger service) and brought it fame with 5 F2a's for semistreamliners on short hauls. In her twilight years the 3004 hauls RDC's and an overflow coach out of Louiseville, Que., for Montreal.

Jim Shaughnessy.

CANADIAN PACIFIC F2a JUBILEE (4-4-4) FOR PASSENGER SERVICE

Series	3000-3004	Weight on lead truck	68,000 lbs.
Cylinders, diameter and stroke	17¼ x 28 in.	Weight on trailing truck	74,000 lbs.
Driving wheel diameter	80 in.	Weight of total engine	263,000 lbs.
Boiler pressure	300 lbs. per sq. in.	Weight of tender, loaded	198,500 lbs.
Grate area	55.6 sq. ft.	Tender capacity	12 tons; 7000 imp. gals.
Evaporative heating surface	2833 sq. ft.	Tractive force, engine	26,500 lbs.
Superheater heating surface	1100 sq. ft.	Builder, date	Montreal, 1936
Weight on drivers	121,000 lbs.		

CANADIAN PACIFIC G3g PACIFIC (4-6-2) FOR PASSENGER SERVICE

Series	2378-2417	Weight on lead truck	58,000 lbs.
Cylinders, diameter and stroke	22 x 30 in.	Weight on trailing truck	64,600 lbs.
Driving wheel diameter	75 in.	Weight of total engine	323,000 lbs.
Boiler pressure	275 lbs. per sq. in.	Weight of tender, 2/3 load	196,760 lbs.
Grate area	65 sq. ft.	Tender capacity	18 tons; 10,000 imp. gals.
Evaporative heating surface	3172 sq. ft.	Tractive force, engine	45,000 lbs.
Superheater heating surface	1475 sq. ft.	Builder, date	Canadian, 1942-1943
Weight on drivers	199,600 lbs.		

J. J. Young Jr.

WHAT THE DOCTOR ORDERED

When the "longest railway in North America" was formed in 1922, a crying need was created for a universal locomotive — one suitable for long-haul passenger or freight assignments, yet easy on the axle loadings. The 4-8-4 filled the bill for Canadian National; between them, CNR and subsidiary GTW bought more than 200. U-2-b 6247 accelerates 18 cars of fans out of Merritton, Ont.

SPIFFY AND SERVICEABLE

The locomotive student got a notebookful of individualism in 1944 when Montreal delivered the 10 U-1-f Mountains to Canadian National. The green 4-8-2's with skirted running boards, single domes, and flanged stacks represented one of the CNR's few departures from utilitarian design. Cone-nosed 6068 flashes across a bridge near Bronte, Ont., with No. 17, the *Inter-City Limited.*

Jim Shaughnessy.

CANADIAN NATIONAL
U-1-f MOUNTAIN (4-8-2) FOR PASSENGER SERVICE

Series	6060-6079
Cylinders, diameter and stroke	24 x 30 in.
Cylinders, diameter an dstroke	24 x 30 in.
Driving wheel diameter	73 in.
Boiler pressure	260 lbs. per sq. in.
Grate area	70.2 sq. ft.
Evaporative heating surface	3584 sq. ft.
Superheater heating surface	1570 sq. ft.
Weight on drivers	236,960 lbs.
Weight on lead truck	57,740 lbs.
Weight on trailing truck	61,000 lbs.
Weight of total engine	355,700 lbs.
Weight of tender, loaded	281,840 lbs.
Tender capacity	18 tons; 11,700 imp. gals.
Tractive force, engine	52,500 gals.
Builder, date	Montreal, 1944

CANADIAN NATIONAL U-2-h NORTHERN (4-8-4) FOR DUAL SERVICE

Series	6235-6259
Cylinders, diameter and stroke	25½ x 30 in.
Driving wheel diameter	73 in.
Boiler pressure	250 lbs. per sq. in.
Grate area	84.3 sq. ft.
Evaporative heating surface	4080 sq. ft.
Superheater heating surface	1835 sq. ft.
Weight on drivers	246,100 lbs.
Weight on lead truck	67,300 lbs.
Weight on trailing truck	86,900 lbs.
Weight of total engine	400,300 lbs.
Weight of tender, loaded	278,150 lbs.
Tender capacity	18 tons; 11,600 imp. gals.
Tractive force, engine	56,800 lbs.
Builder, date	Montreal, 1943

LARGEST AND LAST OF CPR STEAM

Canadian Pacific began buying 2-10-4's in 1929. The last member of a 36-engine fleet, delivered 20 years later, was also the last new steam locomotive purchased by CPR. Selkirk 5928 is in her namesake country as she moves the *Dominion*, train No. 3, away from the Lake Louise station and west toward Kicking Horse Pass. Nobody but CPR ever dared streamline so large a locomotive.

Bruce D. Fales.

BY ROYAL PERMISSION

Following a Canadian tour by King George VI and Queen Elizabeth in 1939 (in which Canadian Pacific 4-6-4's 2850 and 2851 made the 3224-mile Quebec-Vancouver haul without change), permission was asked and granted to name the streamlined engines Royal Hudsons. In the course of their careers the H1d and H1e handled tonnage as well as passengers. Train 43 nears Regina, Sask.

Robert Hale.

CANADIAN PACIFIC H1e ROYAL HUDSON (4-6-4) FOR PASSENGER SERVICE

Series	2860-2864
Cylinders, diameter and stroke	22 x 30 in.
Driving wheel diameter	75 in.
Boiler pressure	275 lbs. per sq. in.
Grate area	80.8 sq. ft.
Evaporative heating surface	3791 sq. ft.
Superheater heating surface	1542 sq. ft.
Weight on drivers	185,800 lbs.
Weight on lead truck	63,900 lbs.
Weight on trailing truck	115,700 lbs.
Weight of total engine	365,400 lbs.
Weight of tender, 2/3 load	229,600 lbs.
Tender capacity	4500 imp. gals. oil; 12,000 imp. gals. water
Tractive force, engine	45,300 lbs.
Tractive force, booster	12,000 lbs.
Builder, date	Montreal, 1940

CANADIAN PACIFIC T1b SELKIRK (2-10-4) FOR FREIGHT SERVICE

Series	5920-5929
Cylinders, diameter and stroke	25 x 32 in.
Driving wheel diameter	63 in.
Boiler pressure	285 lbs. per sq. in.
Grate area	93.5 sq. ft.
Evaporative heating surface	5054 sq. ft.
Superheater heating surface	2032 sq. ft.
Weight on drivers	309,900 lbs.
Weight on lead truck	35,500 lbs.
Weight on trailing truck	101,600 lbs.
Weight of total engine	447,000 lbs.
Weight of tender	281,000 lbs.
Tender capacity	4500 imp. gals. oil; 12,000 imp. gals. water
Tractive force, engine	76,905 lbs.
Tractive force, booster	12,000 lbs.
Builder, date	Montreal, 1938

TO BEGIN POSTWAR MODERNIZATION

Once V-J Day eased equipment supply, the National Railways of Mexico embarked upon an extensive and systemwide overhaul. Alco and Baldwin supplied 32 4-8-4's (known as "Niagras" south of the border) of the QR-1 class, and these were employed on freight runs out of Mexico City to Nuevo Laredo, Aguascalientes, and Guadalajara. The fireman of 3043 examines slipperiness.

CHESAPEAKE IN APPEARANCE . . .

NdeM's 3-foot-gauge Mexico City-Toluca line attains the highest point reached by rail in Mexico, passes through 2 tunnels, and possesses grades of up to 4.8 per cent and maximum curvature of 15 degrees. Following a Coverdale & Colpitts recommendation, 10 simple articulated 2-6-6-2's were purchased between 1928 and 1937. No. 369, in Buenavista Station, Mexico City, looks C&O-ish.

C. W. Witbeck.
L. T. Haug; collection of Everett L. DeGolyer Jr.

Rail Photo Service.

FERROCARRILES NACIONALES DE MEXICO HR-01 2-6-6-2 (3-FOOT GAUGE) FOR FREIGHT SERVICE

Series	369-370
Cylinders, diameter and stroke	15 x 22 in. (4)
Driving wheel diameter	37 in.
Boiler pressure	210 lbs. per sq. in.
Grate area	52.5 sq. ft.
Evaporative heating surface	2449 sq. ft.
Superheater heating surface	680 sq. ft.
Weight on drivers	172,500 lbs.
Weight on lead truck	19,500 lbs.
Weight on trailing truck	24,000 lbs.
Weight of total engine	216,000 lbs.
Weight of tender	110,000 lbs.
Tender capacity	3000 gals. oil; 4500 gals. water
Tractive force, engine	41,100 lbs.
Builder, date	American, 1937

FERROCARRILES NACIONALES DE MEXICO QR-1 NIAGRA (4-8-4) FOR DUAL SERVICE

Series	3033-3048
Cylinders, diameter and stroke	25 x 30 in.
Driving wheel diameter	70 in.
Boiler pressure	250 lbs. per sq. in.
Grate area	77.3 sq. ft.
Evaporative heating surface	4185 sq. ft.
Superheater heating surface	1721 sq. ft.
Weight on drivers	240,000 lbs.
Weight of total engine	387,000 lbs.
Weight of tender, loaded	245,000 lbs.
Tender capacity	6000 gals. oil; 15,000 gals. water
Tractive force, engine	57,000 lbs.
Builder, date	Baldwin, 1946

Collection of H. L. Broadbelt.

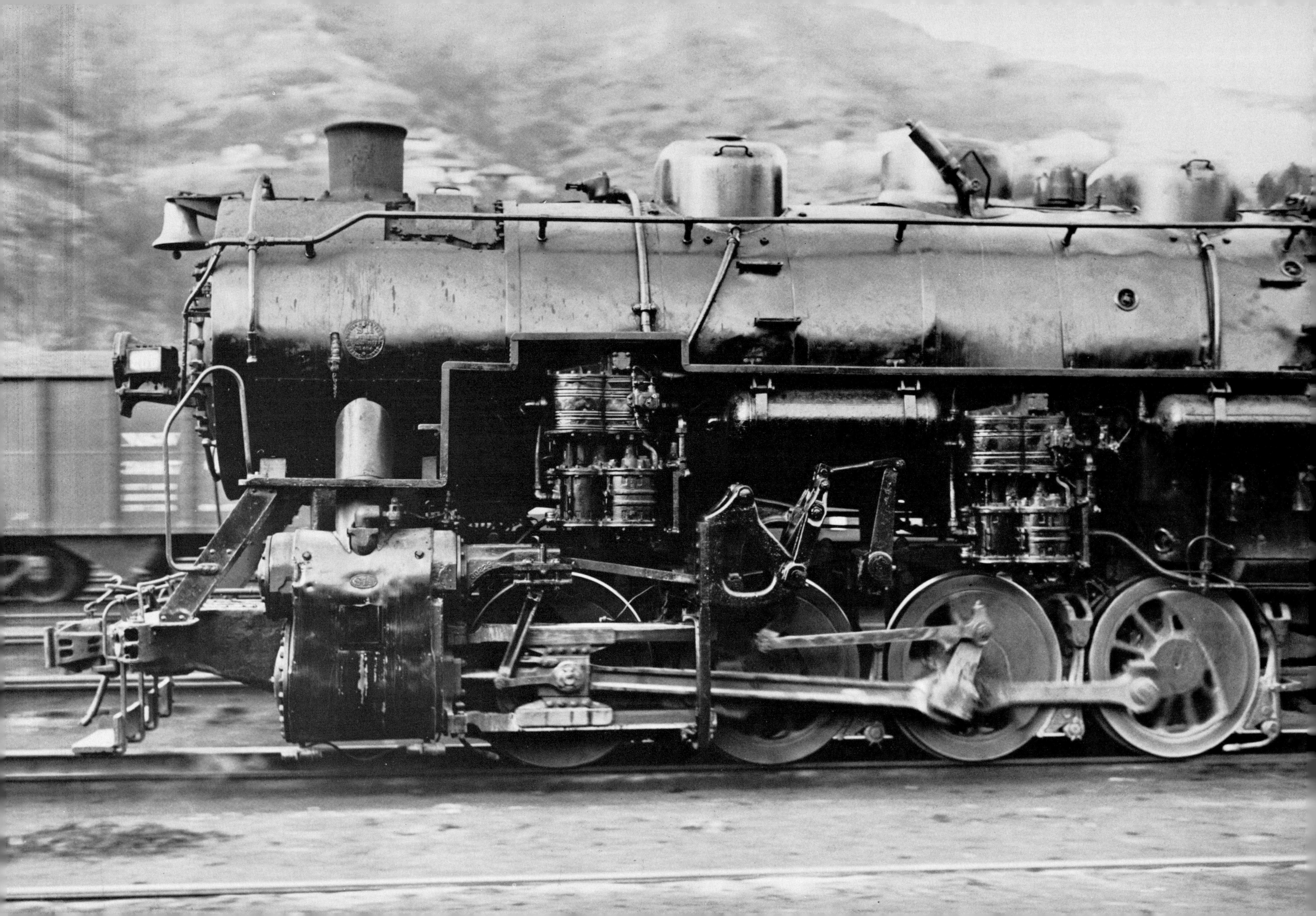

LEST WE FORGET

. . . not quite all the builders' art was lavished upon road power

OF all the neglected wallflowers of modern steam locomotive practice, the switcher was the most forlorn. It was all very well for builders such as Lima to claim that a new shifter could pay for herself out of savings within five years, or for economists to question the efficiency of yard engines produced by the simple expedient of denuding 2-8-0's of their pony trucks. Such talk, however justified, was academic after 1930. During the dark days capital was as scarce as carloadings, and it was patently easier to convince a banker (or a Jesse Jones) of the need for new power on the basis of a reduction in train-miles than on the claim of a percentage reduction in fuel expense per yard locomotive-hour.

In retrospect one wonders how a majority of roads ever got their cars classified. Santa Fe never owned a modern steam switcher; Union Pacific proceeded no further than an 0-6-0 order of 1921. For virtually *all* roads the final evolution of yard engine design occurred during World War I with the introduction of standardized 6- and 8-coupled engines under U.S.R.A. auspices. The road that ventured beyond these two blueprints was the exception (Katy's last new steam engines were big-boilered Lima 0-8-0's which doubled as transfer power).

Well, then, if switching locomotive construction was virtually abandoned after 1930, who *did* buy? Quite simply, those who had no other choice. The Little Giant, Pittsburgh & Lake Erie, kept on buying 0-8-0's clear through World War II, but the inherent economy of the steel hauler is such that it employs more yard crews than road. For P&LE (and for Big Steel's own captive switching and terminal lines, including Aliquippa & Southern; Conemaugh & Black Lick; and Union) there was no ready supply of Consolidations bumped off the main line and available to the yardmaster after amputation of the lead axle. It's understandable that the one new wheel arrangement introduced for nonroad service should have been credited to Union. Acting as a link between Bessemer & Lake Erie and Pittsburgh's mills, Union had a problem prior to 1936: on a transfer run of 80 ore loads an 0-6-0 required assistance on a pair of 2.5 per cent grades. However, a larger locomotive that could increase the number of cars and dispense with the helpers could not exceed 70 feet worth of engine and tender wheelbase. Baldwin's answer was the stoker-fired Union type, an 0-10-2 that could dig in with the tractive effort of a Texas (plus the output of the tender booster) and yet maintain almost 40,000 pounds tractive effort at its maximum speed of 35 miles per hour or so. Horsepower: 3660. In common with such other U. S. Steel engines as Elgin Mikes and Bessemer 2-10-4's, the Unions found a second home on Missabe.

The straight 10-coupled switcher and/or transfer engine — the 0-10-0 — came along early in the century but only two or so roads did much with it in steam's latter days. Alcoa's Alton & Southern was responsible for a novel and handsome engine of the type in 1931, however. A&S's 0-10-0 No. 14 employed so much aluminum in her economy (*e.g.*, in boiler and cylinder jacketing, rods and valve motion, tender booster components, bell, number plate, headlight bracket) that a 10 per cent saving on weight, together with improved counterbalancing of reciprocating parts, was achieved.

If most roads had ceased to buy brand-new switchers when cash ran short, then it might follow that those who continued in the chips (*i.e.*, the Pocahontas coal haulers) had rostered fleets of up-to-date yard engines. Which was certainly the case. Chesapeake & Ohio never ran short on modern 0-8-0's. Just after World War I Chessie tried but never repeated a handful of 10-coupled shifters, thereafter ordered 0-8-0's right through 1930, again in World War II, and still again — 30 engines to be exact — *after* the war. The railroad world, long before committed to diesels, gasped. *New steam shifters in 1948?* One year later Chessie found itself in the camp of its critics, a circumstance which prompted an even more astounding page of locomotive lore. C&O was able to dispose of its brand-new Baldwins, not to the junkers but to a couple of coal-field neighbors: Norfolk & Western and Virginian. Until then N&W, still in no mood to dieselize, had been assigning lighter terminal duties to an aging fleet of turn-of-the-century Consolidations and 4-8-0's. The C&O engines, purchased at bargain-basement prices, arrived in the nick of time. Their performance was so exemplary that the railway enlarged the class by constructing sister engines from scratch in its Roanoke Shops until there were 70 S1's on the roster — 20 secondhanders and 50 homemades. Which brought about a strange and perchance righteous historical footnote: the last reciprocating steam locomotive fabricated new for a U. S. common carrier was not an articulated at all or even a 4-8-4, but an 0-8-0. To be exact, N&W S1a No. 244, Roanoke Shops, 1953.

And rewarding it was in the dieselized 1950's to escape to Roanoke, to hurry through the station to the moving stairway, thence to trackside, and then to walk up the platform . . . toward the whine of a turbogenerator atop an S1's boiler as the 0-8-0 waited to work the head-end cars of No. 17.

J. J. Young Jr.

Homemade Norfolk & Western S1a 0-8-0 No. 206 pedals through the yard at Williamson, W. Va., to coal classification chores. Photo: Philip A. Weibler.

UNION 301 UNION (0-10-2) FOR SWITCHING SERVICE

Series	301-305
Cylinders, diameter and stroke	28 x 32 in.
Driving wheel diameter	61 in.
Boiler pressure	260 lbs. per sq. in.
Grate area	85.2 sq. ft.
Evaporative heating surface	4808.2 sq. ft.
Superheater heating surface	1389 sq. ft.
Weight on drivers	343,930 lbs.
Weight on trailing truck	60,430 lbs.
Weight of total engine	404,360 lbs.
Weight of tender	240,000 lbs.
Tender capacity	14 tons; 12,000 gals.
Tractive force, engine	90,900 lbs.
Tractive force, booster	17,150 lbs.
Builder, date	Baldwin, 1936

CONEMAUGH & BLACK LICK 0-8-0 FOR SWITCHING SERVICE

Series	47-48	Weight on drivers	230,800 lbs.
Cylinders, diameter and stroke	25 x 28 in.	Weight of total engine	230,800 lbs.
Driving wheel diameter	51 in.	Weight of tender, 2/3 load	122,017 lbs.
Boiler pressure	190 lbs. per sq. in.	Tender capacity	11 tons; 8200 gals.
Grate area	47 sq. ft.	Tractive force, engine	55,400 lbs.
Evaporative heating surface	2777 sq. ft.	Builder, date	Lima, 1941
Superheater heating surface	639 sq. ft.		

PITTSBURGH & LAKE ERIE U-3l 0-8-0 FOR SWITCHING SERVICE

Series	8050-8074	Weight on drivers	234,000 lbs.
Cylinders, diameter and stroke	25 x 28 in.	Weight of total engine	234,000 lbs.
Driving wheel diameter	52 in.	Weight of tender, loaded	188,000 lbs.
Boiler pressure	190 lbs. per sq. in.	Tender capacity	16 tons; 10,000 gals.
Grate area	47 sq. ft.	Tractive force, engine	54,350 lbs.
Evaporative heating surface	2777 sq. ft.	Builder, date	American, 1944
Superheater heating surface	639 sq. ft.		

NORFOLK & WESTERN S1a 0-8-0 FOR SWITCHING SERVICE

Series	200-214
Cylinders, diameter and stroke	25 x 28 in.
Driving wheel diameter	52 in.
Boiler pressure	220 lbs. per sq. in.
Grate area	46.9 sq. ft.
Evaporative heating surface	2570 sq. ft.
Superheater heating surface	637 sq. ft.
Superheater heating surface	637 sq. ft.
Weight on drivers	247,000 lbs.
Weight of total engine	247,000 lbs.
Weight of tender, loaded	212,270 lbs.
Tender capacity	15 tons; 13,000 gals.
Tractive force, engine	62,932 lbs.
Builder, date	Roanoke, 1951

ALTON & SOUTHERN 0-10-0 FOR SWITCHING SERVICE

Number	14
Cylinders, diameter and stroke	28 x 30 in.
Driving wheel diameter	57 in.
Boiler pressure	230 lbs. per sq. in.
Grate area	80 sq. ft.
Evaporative heating surface	4009 sq. ft.
Superheater heating surface	1116 sq. ft.
Weight on drivers	320,840 lbs.
Weight of total engine	320,840 lbs.
Weight of tender, loaded	238,960 lbs.
Tender capacity	18 tons; 12,000 gals.
Tractive force, engine	80,500 lbs.
Tractive force, booster	15,800 lbs.
Builder, date	Baldwin, 1931

Collection of H. L. Broadbelt.

Postscript

ON the eve of dieselization even the converts couldn't quite visualize railroading without the steam locomotive. The diesel at first was pegged for specific slots — for yard duty in smoke-conscious cities, then for lightweight streamliners, later for freight hauls in mountainous or very dry country. Yet the implication of steam and diesel working side by side was akin to lambs and lions inhabiting the same cage. But the faithful continued understandably to clutch at straws. The Government would buy up modern steam locomotives and store them against the chance of a national emergency . . . granger roads would call out steam during the annual wheat rush rather than buy expensive diesels for seasonal service . . . steam would always be operated in commutation and other low utilization services — one by one, the heartening, plausible excuses were published and accepted at face value. And subsequently demolished. Finally even Norfolk & Western surrendered.

There were unexpected oases in the diesel desert. Both Illinois Central and Union Pacific had tinkered with the diesel in its infancy, yet both were dispatching tonnage behind steam years after Chesapeake & Ohio had scrapped steam power built in 1948. Baltimore & Ohio had halted steam power research long before World War II, had purchased the first non-articulated passenger diesel and some of the pioneer freight units, but still contrived to darken the Indiana sky with the exhaust of double-slotted 4-8-2's after Pennsy had gone diesel *in toto*. During one celebrated diesel shortage, Central's 4-6-4's and 4-8-2's impishly steamed it up in the formerly sacred and electrified premises of Cleveland Union Terminal. Regardless of where or why a column of locomotive smoke rose in the waning 1950's, however, the faithful were abroad in the land to spot it and to come to renew old acquaintance.

Nor has the scrap merchant claimed all the diesel's victims. Disgracefully enough, most of the pioneers did get the torch — Lima's A-1 2-8-4, the first Hudson, Timken's *Four Aces*, and far too many more. But at least Espee's last new steam engine, cab-forward 4-8-8-2 No. 4294, is parked forever near the steps of the Sacramento station, and visitors to Ford's Greenfield Museum come upon a Chessie 2-6-6-6 that constitutes a more memorable exhibit than most of them realize. And there are other engines preserved from St. Louis to Pasadena — which is as it should be.

The lasting memorial to the modern steam locomotive lies not in statistics or upon the printed page or even on museum grounds. It is in the hearts and minds of all who saw and heard the big power in action. The locomotives included in this volume are forever rolling for these people, whether they were participants or among the audience.

They will not forget.

For them the time will always remain as it once was: steam's finest hour. **I**

Rebuilt Illinois Central 2-8-2 1563, her boiler uncommonly cold, is cloaked in ice while awaiting scrap on a siding in Champaign, Ill., in January 1959.

J. P. Lamb Jr.

U.P.
4024
25,000 GALLONS
DUPONT 11-28-44